Springer Praxis Books
Popular Science

This book series presents the whole spectrum of Earth Sciences, Astronautics and Space Exploration. Practitioners will find exact science and complex engineering solutions explained scientifically correct but easy to understand. Various subseries help to differentiate between the scientific areas of Springer Praxis books and to make selected professional information accessible for you. The Springer Praxis Popular Science series contains fascinating stories from around the world and across many different disciplines. The titles in this series are written with the educated lay reader in mind, approaching nitty-gritty science in an engaging, yet digestible way. Authored by active scholars, researchers, and industry professionals, the books herein offer far-ranging and unique perspectives, exploring realms as distant as Antarctica or as abstract as consciousness itself, as modern as the Information Age or as old our planet Earth. The books are illustrative in their approach and feature essential mathematics only where necessary. They are a perfect read for those with a curious mind who wish to expand their understanding of the vast world of science.

Maurizio Agrò

Music and Astronomy

From Pythagoras to Steven Spielberg

Maurizio Agrò
Siracusa, Italy

ISBN 978-3-031-41523-4   ISBN 978-3-031-41524-1 (eBook)
https://doi.org/10.1007/978-3-031-41524-1

Translation from the Italian language edition: "Musica e Astronomia, da Pitagora a Steven Spielberg" by Maurizio Agrò, © The Author(s) 2022. Published by Discovery Edition. All Rights Reserved.

© The Editor(s) (if applicable) and The Author(s), under exclusive license to Springer Nature Switzerland AG 2023
This work is subject to copyright. All rights are solely and exclusively licensed by the Publisher, whether the whole or part of the material is concerned, specifically the rights of reprinting, reuse of illustrations, recitation, broadcasting, reproduction on microfilms or in any other physical way, and transmission or information storage and retrieval, electronic adaptation, computer software, or by similar or dissimilar methodology now known or hereafter developed.

The use of general descriptive names, registered names, trademarks, service marks, etc. in this publication does not imply, even in the absence of a specific statement, that such names are exempt from the relevant protective laws and regulations and therefore free for general use.

The publisher, the authors, and the editors are safe to assume that the advice and information in this book are believed to be true and accurate at the date of publication. Neither the publisher nor the authors or the editors give a warranty, expressed or implied, with respect to the material contained herein or for any errors or omissions that may have been made. The publisher remains neutral with regard to jurisdictional claims in published maps and institutional affiliations.

This Springer imprint is published by the registered company Springer Nature Switzerland AG
The registered company address is: Gewerbestrasse 11, 6330 Cham, Switzerland

Paper in this product is recyclable.

Foreword

If you want, you could do as I did. I started reading the book—and you shouldn't, of course—from the last chapter, "Close Encounters of the Third Kind" (the title of a 1970s science fiction movie). I did so because the title literally seduced me the way Devil's Mountain in Wyoming attracted Roy. The mysterious mountain was the place where the protagonist would have an encounter with aliens. The first approach with those beings was through five musical notes that for years (and even now) resonated in my mind and I think also in the minds of those who saw that film. The five notes that were used to find a way to communicate with beings from other worlds were not chosen at random. Those five notes sum up all the philosophical, anthropological, scientific, and musical discourse that has been going on over the centuries starting with the need to understand what music is and what connection it has with the Universe.

But after reading only a few lines of the last chapter, it was not long before I felt the urge to start from the beginning, as I should have done. The result was a journey between music and celestial spheres. Needless to mention here that music is based on the harmony of notes, and needless to mention that the Universe is based on physical rules that have their own harmony, but it is curious that so many musicians with knowledge of physics and astrophysics have been unable to resist the temptation to transform the patterns of radiation from space into musical sounds, even creating genuine pieces of music.

This common thread between music and astronomy was already perceived, perhaps even more so than today, by ancient philosophers, emphasizing this link in medieval times with the *quadrivium*, which brought together the four disciplines making up the mathematical arts: arithmetic, geometry, astronomy, and music. And so it is right up to the present day, where eminent astrophysicists have likened the residual signal of the Big Bang to a cry, to the weeping of a newborn baby, the sound of two colliding black holes to a chirp, and even the regular emissions of a pulsar to a song.

In the pages that follow you will find an unparalleled discussion and detailed comparisons of the common features that unite music with astronomy and astronomers with music.

This is a rigorous book, but Maurizio Agrò is always ready and willing to explain the phenomena and the people encountered along the way, with a host of anecdotes that bring so much more to a subject that is already of infinite scope: music and astronomy.

Milan, Italy Luigi Bignami
June 2023

Preface

My urge to investigate the music of the celestial spheres first started about a decade ago during some research on the relationship between musicians and scientists. In this respect, a meeting in 2014 with Dr. Giuseppe Cutispoto, an astronomer at the Catania Astrophysical Observatory and head of outreach at the Istituto Nazionale di Astrofisica (INAF), was all important.

We first collaborated on an astronomy project in middle school, and later I had the pleasure of attending his seminars. One of his lectures was entitled *The Human Exploration of Space: From Gagarin to the Conquest of Mars*. I was bowled over. All the ideas and research carried out so far suddenly took shape and resulted in a seminar proposal that Dr. Cutispoto enthusiastically accepted, even giving me the opportunity to hold two seminars at the Astrophysical Observatory of Catania. The title was "The Music of the Spheres."

In September 2017, during the conference "The Future of the Universe," I met Dr. Angelo Adamo, a musician and astronomer, who was giving his contribution "ConcertFerenza": *Getting in tune with the cosmos: how visual observation has acquainted us with its voice*. A deeper study of the relationship between music and the cosmos was in order. The way Adamo presented things was clear and magical. You were transported into the very heart of the forming Universe and could hear its voice.

While writing this book, I came across a song by Dik Dik called *Help me*, which raised several questions for me. I had the opportunity to chat with Pietruccio Montalbetti, founder of the historic group and author of the song, who immediately accepted to open a dialogue on the subject. In this volume, I report part of our conversation.

This book grew out of the trust Dr. Cutispoto placed in me. It may not seem like much, but gaining approval from an authoritative source is a major boost to continuing one's work. My warmest thanks go to him, to Dr. Angelo Adamo, to Pietruccio Montalbetti, and to all the authors of the books I mention as I go along, because each with their own professionalism has added a piece to a puzzle that I began to compose ten years ago.

Syracuse, Italy
June 2023

Maurizio Agrò

Contents

1	**Introduction**	1
	References	2
2	**The Universe Born of Sound**	3
	References	10
3	**Everything Is Number. Pythagoras and the Pythagoreans**	13
	References	25
4	**The Music of the Spheres in Kepler's Cosmos**	27
	References	34
5	**Scientist Musicians**	35
	References	40
6	**From Newton to Einstein**	41
	References	49
7	**The Music of the Cosmos**	51
	References	55
8	**The Universe Is Teeming with Aliens**	57
	References	67

9	The Songs of the Space Race	69
	References	80
10	Space, the Music of the Spheres, and Movies	81
	References	85
11	Close Encounters of the Third Kind	87
	References	99
12	Conclusion	101
	References	103
Appendix		105
Index		107

1

Introduction

Painters, musicians, filmmakers, and storytellers are all out to measure up the world with their films, their colours, their musical notes, and their words: all descriptions which, taken together with their complementary scientific accounts, most certainly do not weaken our understanding of the cosmos, but rather enrich it with things that measurement – understood as a purely physical process – could never give us [1].

I am pleased to begin this discussion by quoting the astronomer and musician Angelo Adamo because these words sum up so well the idea that science cannot be separated off from art, while art finds inspiration in our knowledge of Nature. The need to describe is the fundamental starting point for understanding our world, and only through the work of intellectuals, scientists, artists, and musicians is it possible to have a clear vision of who we are and what we are here for. Man was born to measure things, to be a witness to creation, to be a modern *aoidos*, able to explain and predict.

Through the concept of the music of the celestial spheres, we tried to explain the workings of the world in which we are immersed, that is, our universe. But first, we need to answer Carl Dahlhaus's question about what music is and about the indefiniteness of the concept of "music" [2]. In fact, an unambiguous definition has never been found, so the term "music" has become synonymous with many different expressions, but it is certainly well suited to the description of a harmonious, well ordered universe, that is, a Cosmos. On the other hand, we can adopt the opposite reasoning: music as order, as harmony, becomes a representation of the Universe itself. It was no coincidence that ancient peoples considered music, not as an art, but as a science, and it was no coincidence that Ptolemy would speak of harmonic science.

The discovery that our world can be represented by precise mathematical models brings us to consider the existence of a *vis harmonica* that regulates all things, and the only discipline able to represent this force is music. Therefore, this "music" can be used for measurement because it involves space, through the linear way it is written and the linearity of its melodies, and time, through the duration of its notes. Moreover, it is the only discipline that can be associated with the concept of four-dimensional space-time. In fact, in music, we can identify three space dimensions, namely the melodic line as x-axis, the harmonic line as y-axis, and the orchestral line as z-axis, and one time dimension. According to Einstein's theory of relativity, space and time are linked, and this is also true for music, because, in a sonata, if I increase the playing time, the melodic line contracts. This is what happens when you travel at speeds close to the speed of light, so music is well suited to describing certain phenomena that occur in our universe. Scientists commonly use musical terms and examples to explain our world. One only has to think of string theory, inspired by the vibration of a violin string. Beyond our earthly music in which sound itself is the creator, there is also a type of music inspired by science and by space travel. Here, government agencies such as NASA respond to the music of the universe with a *musica instrumentalis*, launching probes with musical recordings and bringing musical instruments aboard spacecraft as if to re-establish that ancient connection between humanity and the cosmos, mentioned by Pythagoras, in which music is the means of communication. Steven Spielberg embodied exactly this spirit when he shot "Close Encounter of the Third Kind", and in the light of the reasoning exposed in these pages, it will not be difficult to understand how close the link is between astrophysics and music even today.

References

1. A. Adamo, *Pianeti tra le note* (Springer-Verlag Italia, Milan, 2010)
2. C. Dahlhaus, H.H. Eggebrecht, *Was ist Musik?* (Heinrichshofen, Bologna, 1985)

2

The Universe Born of Sound

Whenever the genesis of the world is described with sufficient precision, an acoustic element intervenes at the decisive moment of the action. [1]

With these words Marius Schneider opens the first chapter of his essay *Primitive Music*, exploring those cultures that place the myth of a creative sound at the center of their cosmogony.

In order to answer the question "Where do we come from?" we must examine human history from the perspective of the beliefs and mythology of ancient peoples, not considering these narratives as a product of pure human imagination, but viewing them rather as a way of passing on atavistic knowledge using the expressive means appropriate to their historical period. There is a widely held idea that, when mythological events are put forward as leading to the creation of the world, this is a consequence of scientific ignorance, yet there is always some truth in these teachings, described as they are through the eyes of those who, for the first time, become aware of their human condition and their place in the World. Clearly, inspiration for these descriptions arises from the surrounding environment, and so it is that we hear of singing caves, speaking, shouting, and shrieking animals, thunder that creates the Earth and sky, and gods that move in the vault of heaven, fighting for man or confronting him. Two basic features are always present among this variety of elements: vibration and propagation. Both are fundamental aspects of any radiative process, referred to in modern times as a wave. So, there can be no emission of sound, no sound wave, unless there is a vibrating body and a medium for the propagation of vibrations. Here, then, a character appears on the scene and performs an action capable of emitting a vibration into the void, and it propagates in the void, giving rise to all the elements that constitute the World. In Wolfgang Petersen's film *The Neverending Story* (1984), the daring warrior Atreyu attempts to save the Kingdom of Fantasia from the advancing Nothing. In this filmic tale, the Nothing symbolizes man's loss of imagination and fantasy. The story is rewritten as Bastian, the protagonist, reads it, and when he shouts out his mother's name, the Kingdom of Fantasia will be saved and the Nothing will be defeated. The valiant Atreyu will return to run through the meadows of a new world, flown over by Falkor, the dragon of fortune. Bastian's mother's name will be the new name of the Childlike Empress, ruler of Fantasia, who will bring a new kingdom into being.

Bastian: *Why is it so dark?*
Childlike Empress: *It is always dark in the beginning.*
Bastian: *What is that?*
Childlike Empress: *One grain of sand. It is all that remains of my vast empire. [...] Fantasia can arise anew from your dreams and wishes [...] the more wishes you make, the more magnificent Fantasia will become.*
Bastian: *Then my first wish is...*

The story told by Michel Ende, published as a novel in 1981, is not unlike the stories handed down by ancient peoples about birth in our world: a hiss, a breath, or a voice become the creators of Creation.

The term "cosmogony" refers to the set of theories and myths that explain the birth of the cosmos and the ordering of the Universe in order to reveal the creation of the world and humankind. All peoples possess one or more cosmogonies that carry with them the imprint of each society's cultural identity [2].

In Mayan mythology, the creator god Hurakán, flying in primordial darkness above an expanse of water, uttered the word "Earth," and a mass of solid Earth emerged from the waters. The Mascalero Apaches of North America tell that the world was created by the god Libayè with song and dance, while the Australian god Pundjel created two men from clay and pieces of bark, blowing on them and thereby bringing them to life. In ancient Egypt, the god Khum, after creating man, animated him with a puff of his own breath, and during the Memphite age, the Egyptian god of the city of Memphis, Ptah, created the cosmos through speech. The account is well documented in the transcription of the *Shabaka Stone*, which is part of the so-called "Pyramid Texts." The text reads:

> *[...] He formed all the gods, including Aton. Every divine word had existence, by the thought of the heart and the command of the tongue. [...] It was he who created quantities and qualities; he by his word made all foods and all offerings; he made*

what is loved and what is hated; he gave life to the peaceful and death to the guilty. [...] [3]

Another creation myth is found in the Bremner–Rhind Papyrus, in which it is once again the voice of God that generates the Universe:

[...] The Lord of All said, after coming into being: I am he who came into being as Khepri. When I came into being, being itself came into being, and all beings came into being after I came into being. Many were the beings that came into being from my mouth... [...] [4]

And in the Leiden papyrus we can read:

[...] And God laughed seven times: Ha, Ha, Ha, Ha, Ha, Ha, Ha, and when God laughed, there arose seven gods [...] [5]

It would not be wrong to compare this energetic laughter to the Christian creation myth in Genesis, which tells how God made the world in seven days.

In the Polynesian creation myth, in the beginning there was only Po, or chaos, still and silent; light and day arose from its motion and groaning.

The Makiritare Indians, originally from Venezuela, tell how the man and woman danced and sang inside a large shining egg, and God, as he broke the egg, sang the words: *I break this egg and the woman is born and the man is born.* In this myth we find two very interesting elements: the word/sound and the egg. Many cosmogonies attribute the birth of the World and the Universe to a primordial Egg, the cosmic Egg from which everything sprang. In fact, the egg myth can be traced back to the primordial soup theory, and in some ways, we can read the birth of our universe as coming from a bubble in which all four fundamental forces were united—the strong nuclear force, weak nuclear force, electromagnetic force, and gravitational force, that is, the laws of Nature.

In 1927, the astronomer George Lemaître concluded that the universe originated from a "*cosmic egg exploding at the moment of creation*" [6].

The Indian creation myth recounted in the Satapatha-Brahmana document tells of a golden egg produced from hot waters, the only things existing at the beginning of the Universe. After one year, Prajapati emerged from the egg and the following year "*[...] tried to speak. He said 'bhuh!' and this became the Earth; 'bhuvah,' and this became the air; 'svah,' and this became our sky*" [6]. Actually, we do not know the true sound of creation. However, we can trace the use of the word, but not so much from the onomatopoeic sound as from the words "And God said."

The best known myth is certainly the Creation as narrated in the Bible:

- God said, "Let there be light!" And the light was.
- God said, "Let there be a firmament in the midst of the waters [...]"
- God said, "[...] Let the waters be gathered in one place."
- God said, "Let the earth produce shoots [...]"
- God said, "Let there be lights in the firmament of heaven [...]"
- God said, "Let the waters teem with living things [...]"
- God said, "Let the earth produce living beings [...]"
- God said, "Let us make man in our image [...]"

So far God created the earth the sky and living beings, but the Lord noticed that no one worked the earth, so "[...] the Lord God molded man with dust from the ground and breathed into his nostrils, and man became a living being" [7].

In this narrative, it is clear how God uses the Word to found the World and Breath to give Man the essence of being. Once again then, sound in its phonetic form is responsible for the beginning of all things.

Another Christian-based creation narrative goes as follows:

In the beginning was the Word, and the Word was with God, and the Word was God. He was in the beginning with God: everything was made through him, and without him nothing was made of all that exists. [8]

The Word is thus the divine Logos which created the Universe.

In Faust's meditation on the narrative of the Gospel according to John, he cannot accept the translation of Logos into Word:

[...] It is written, 'In the beginning was the Word!' Here already I get stuck. [...] To place the word so high I cannot. I have to translate it some other way. [...] It should be: In the beginning was strength! [...] Suddenly I see clearly and write with confidence: In the beginning was the act! [...]. [9]

If the idea of "force" can lead us to think that energy is at the origin of everything, it is legitimate to consider the laws of nature as a priority in the construction of the world. The need to go even further by using the word "act" presupposes that at the beginning there is the *doing* or the action on which both nature and knowledge depend.

From a philosophical point of view, it is therefore possible to argue about the Logos/God duality from different points of view, but this kind of speculation is not interesting for our purposes here. In fact, beyond the interpretative arbitrariness of the various myths, there are two common features in almost all cosmogonies: place (Cosmic Egg, Earth, Waters, Heavens) and sound (word, breath, thunder).

In the main cosmogonies, there is always a creative force with anthropomorphic traits that may manifest itself in the form of a god, father, primordial king, or earth goddess, but generally as a deity symbolizing some natural phenomenon, including thunder [10]. Among American Indians, the Lakota tell how the Wakinyan, allies of the Great Spirit Wakan Tanka, generate life with thunder, which symbolizes the divine voice:

Creation begins when the star descends to the Earth and does not burn it, at the very moment the Creator spirit landed on the Earth, the Thunder which is the Sound of the Creator was heard for the first time.

For the Apache people of North America, it is not known how Usen created the universe, but when he made the Earth, he commanded the four spirits Black Water, Black Metal, Black Wind, and Black Thunder, to do it for him. These were thus the demiurges of creation. Black Thunder gave the Earth her hair in the form of grass and trees, while Black Wind gave her breath in the form of euphoria [11].

Cosmogony is often associated with theogony, from which arises some form of religion that employs a "*[...] system of symbols that places man in relation to the universe and allows the multiplicity of phenomena to be given order*" [12]. Religious practices help reduce the sense of helplessness when faced with events that, in their everyday lives, humans cannot control. These practices, along with magical rituals, arise when a society lacks the technical and scientific knowledge to explain certain phenomena.

Comparing the various cosmogonies and creation myths, certain elements can be picked out that may point towards a scientific interpretation. Many narratives refer to mists, darkness, waters, and skies, all factors that are easily recognizable in our modern knowledge of the interstellar medium, namely, space. From photographs taken by the Hubble Space Telescope, we have come to understand that the universe is made up of "dust," itself largely made up of molecular hydrogen, a fundamental constituent of water, which obscures certain regions of the cosmos. The "heavens" are separated from the "waters," and the "mists" (cloud), which are opaque to radiation ("dark"), generate "the Universe" by Thunder or the Word (gravity) (worlds are born from an accretion disk). It is therefore possible that creation myths tell a scientifically plausible story, but told with the tools of the imagination. According to Lévi-Strauss

"[...] although each myth has the appearance of a bizarre tale from which all logic is absent, there are simpler and more intelligible relations among them than the stories told [...]" [13]. In the second half of the twentieth century, new theories were popularized that saw cosmogonies as representing a point of contact between human and extraterrestrial civilizations.

This gave rise to pseudoscientific disciplines bearing names like the ancient astronaut theory, the paleocontact theory, and paleoastronautics. These sprang from the publication of Erich von Däniken's book *"Chariots of the Gods?"* (1968), although in 1957 the Italian author Peter Kolosimo, whose real name was Pier Domenico Colosimo, had published a work called *"The Unknown Planet,"* in which he attempted to explain certain archaeological mysteries in terms of a kind of pseudo-archaeology from space. The most exciting contribution to "heretical" archaeological theories came from Zecharia Sitchin, author of a great many texts on mysterious archaeology and the ancient astronaut theory. Sitchin believed that the birth of the Sumerian culture should be attributed to an alien race called the Anunnaki. The sources of inspiration for his theories are the cuneiform tablets narrating the Sumerian cosmogony and the birth of Man. According to Sitchin, in this ancient cosmology, the Solar System had a tenth planet with an elliptical orbit and a period of revolution of 3600 years. This planet was called Nibiru, and in Babylonian mythology it was associated with the god Marduk. The Anunnaki were supposed to be the inhabitants of Nibiru who had come to Earth to mine gold, a material they needed to protect their planet's atmosphere from destruction. According to Sitchin, they created human beings on Earth from a hybrid of their own DNA and that of early hominids in order to build a workforce for mining. In support of this idea, he mentions a seal in which the Ancient Sumerians regarded Earth as the seventh planet, thus suggesting a view of the Solar System from the outside, whence Pluto, rather than Mercury, becomes the first planet.[1]

According to Sitchin, *The Epic of Atra-Hasis* relates that, upon Anu's return to Nibiru, the Anunnaki worked in the mines of Abzu for "forty numbered periods." The work was strenuous and there was a revolt, a genuine mutiny. Enlil, Anu's son, sent a message to his father asking him to come down to Earth, and assembled a court martial in which his brother Enki, leader of the Abzu, was also present. He had a solution to propose, given also the presence of Sud, sister and chief medical officer:

Let us have her create a Primitive Worker so that he will be the one to carry the yoke…let the Worker perform the toil of the gods, let him be the one to carry the yoke!

Enki suggested using an "already existing being and imprinting on this evolved being the mold of the gods." The goddess Sud purified the "essence" of a young male Anunnaki by introducing it into the egg of a female monkey. When the creature came into the world, Sud exclaimed, "I created him! My hands made him!" The first "Primitive Worker," *Homo sapiens*, had been born [14].

Although Sitchin's account is full of details that may appear more or less plausible, the narrative is undoubtedly fascinating, on par with a science fiction novel.

The poem of Atra-Hasis is the oldest known religious work recounting the Genesis. It comprises three tablets which constitute the historical, archaeological, and scientific reference used by Sitchin to develop his theory. The narrative refers to man's creation and the early conflicts that opposed him to the gods, before the great flood, whose protagonist is Ziusudra, the Noah of the Bible, and other unsuccessful attempts to destroy humankind [15]. Thus, the ancient astronaut theories are based on actual historical and archaeological knowledge, but they attempt to provide an "alternative" explanation that is inconsistent with canonical studies.

Of course, there is no scientific evidence to support these theories, but what is interesting is to witness the emergence of new cosmogenic myths despite current scientific knowledge. Indeed, these new theories have succeeded in originating new, modern religions based on distrust of official science rather than on lack of such knowledge.

However, applying current scientific findings to mythology, it is possible to attempt an explanation for certain phenomena, but the same method may prove ineffective when applied, for example, to the destruction of Sodom and Gomorrah. It is said that the two cities were destroyed by a rain of fire and brimstone falling from the sky, and this depiction is often identified with destruction by nuclear weapons, following Sitchin. However, a more scientific explanation might be found in a purely natural phenomenon, namely the impact of a meteorite.

Myths can therefore be interpreted in different ways and relative to different cultures at different times. This is not to say that mythology is just a fictional tale, but that humans have attempted to pass on knowledge that is not fully understood, but is of considerable importance in that particular culture. According to Malinowski, "*[...] It is not invention, as a novel might be, but it is a living reality, believed to have happened in primeval times, and which endures long enough to influence the world and human destinies*" [16]. Myths thus transform a physical reality into a precise act and reconnect it with a supernatural origin: history is integrated into myths, and myths are situated at the beginning of history. What peoples without writing ask of myths is to explain the origin of the world and their role in societies, and this way of interpreting the past depends on the environment to which one belongs [13]. Lévi-Strauss states that myths reveal the unconscious of a people and expresses symbolically the meaning of such primordial questions as "Where do we come from?", "Who are we?", and "Where are we going?" [10]. For Gianbattista Vico, myth is "*[…] a descriptive form, poetic and figurative, of natural facts and phenomena,*" while De Santillana and Von Dechend consider that "*[…] the complexity of mythological stories is not inferior to that of current scientific concepts; a reading on different planes and levels is needed to understand it more thoroughly*" [17].

In Odifreddi's words, scientific answers to classical theological questions are more complex and less reassuring than those given by mythology [18].

Note

1. The planet Pluto was downgraded to a dwarf planet by the International Astronomical Union in 2006 because of some of its characteristics, such as being the smallest in the Solar System. At the time of writing of Sitchin's book, Pluto was still officially a planet.

References

1. M. Schneider, Primitive Music, in *New Oxford History of Music. 1 Ancient and Oriental Music*, ed. by E. Wells, (Oxford University Press, Oxford, 1957)
2. J. Ries, *Alla ricerca di Dio, la via dell'antropologia religiosa* (Jaca Book, Milano, 2009)
3. F. Bandini, *E fu sera e fu mattina: primo giorno. I miti della creazione tra Eros e Ethos. Quaderni di etnologia e archeologia del sacro* (Alinea Editrice, Florence, 2006)
4. M. Agrò, *L'Antico Egitto e la Musica* (Ananke, Turin, 2009)
5. A. Colimberti (ed.), *Ecologia della musica, saggi sul paesaggio sonoro* (Donzelli Editore, Rome, 2004)
6. P. Blom, *Fracture: Life and Culture in the West, 1918–1938* (Basic Books, New York, 2015)
7. The Holy Bible, Genesis (Nelson Bibles, 2018)
8. The Gospel of John (Baker Academy, 2015)
9. W. Goethe, *Faust* (OUP, Oxford, 2008)
10. C. Rivière, *Introduzione all'antropologia* (Il mulino, Bologna, 1995)
11. W. Pedrotti, *Leggende e miti dei Pellerossa* (Demetra, Verona, 2000)

12. U. Avalle, M. Maranzana, P. Sacchi, *Antropologia Culturale* (Zanichelli, Bologna, 2000)
13. C. Lévi-Strauss, *L'Anthropologie face aux problèmes du monde moderne* (Editions du Seuil, Paris, 2011)
14. Z. Sitchin, *The Wars of Gods and Men* (Avon books, New York, 1985)
15. J.C. Margueron, *La mesopotamia* (Editori Laterza, Rome–Bari, 1993)
16. B. Malinowski, Mito e antropologia, in Mito e filosofia, V. La dimensione pratica del mito, Leonardo Lotito (Bruno Mondadori, Milan, 2003)
17. G. Pigoli, *I dardi di Apollo* (UTET, Turin, 2009)
18. P. Odifreddi, *Il vangelo secondo la scienza* (Einaudi, Milan, 2008)

3

Everything Is Number. Pythagoras and the Pythagoreans

In antiquity, the musical question was considered by many thinkers such as Aristoxenus in the *Elementa Harmonica*, Euclid in *Sectio Canonis*, Ptolemy in the *Harmonics*, and Theon of Smyrna in a treatise on numbers and music. Each linked the problem of musical intervals with numerical ratios and their model of the Cosmos.

In the work *De Placita Philosophorum* by Aetius, a Greek philosopher, it is stated that Pythagoras, recognizing the order of the universe, called it Kosmos [1]. Actually, the term in the cosmological sense is already found in Anaximenes, predating Pythagoras, and in Heraclitus, his contemporary, but we find it again in Empedocles and Parmenides, posthumous to the philosopher of Samos [2].

The Pythagorean school was the first to study astronomy using mathematical methods, although it subsequently became the task of the philosophers of the fourth century BCE to develop the discipline. The concepts of "cosmos" and "harmony" coincide in the Pythagorean understanding of astronomy. For the Pythagoreans, musical phenomena were the counterpart of cosmic harmony realized by the motion of celestial bodies. Pythagoras was not interested in music from a practical point of view, rather he was interested in its scientific nature, becoming the first to discover the relationship between mathematical ratios and musical intervals. Iamblichus' account in *Life of Pythagoras* has become one of the most famous explanations of how Pythagoras came to formulate the first scientific theory of musical ratios:

On one occasion [Pythagoras] [...] passed a blacksmith's workshop and [...] heard hammers beating iron on the anvil and making sounds, all in perfect harmony with each other, except for one combination. In those sounds, Pythagoras recognized the chords of octaves, fifths, and fourths, and noted that the interval between fourths and fifths was in itself dissonant, but suitable for bridging the difference in magnitude between one and the other [...] he understood that the difference in the pitch of the sounds depended on the weight of the hammers and not on the force with which they were beaten, nor on their shape [...] After determining the weight of the hammers with the utmost precision, he went home. Here he fixed a single peg to the corner of two walls [...]; to the peg he attached one after another four strings of equal thickness and tension [...] taking care that the strings were of perfectly equal length. Then plucking the strings two by two alternately, he found the already mentioned chords, one for each pair of strings. Indeed, he understood that the string stretched by the larger weight resonated in an octave relationship with the one stretched by the smaller weight: one had a weight of twelve units and the other of six. He thus showed that the octave is based on a ratio of 2:1, the same ratio as these weights. The string with the largest weight resonated in a fifth chord with the one that carried eight units of weight and was placed next to the string carrying the smallest weight: in this way, he proved that the fifth is based on a ratio of 3:2, the same ratio in which the weights stood. [...] [3].

This account is surely part of a legend that arose to explain the method used by Pythagoras in his experiments. It may therefore be considered that the Pythagoreans used the inductive method when carrying out experiments on the nature of intervals.

Hippasus of Metapontum and Lasus of Hermione used the criterion of varying the velocities of motions. The former made four bronze discs with the same diameter but with proportional thicknesses in order to create consonances, while the latter filled vessels with water in different proportions and by striking them obtained the various intervals: the octave, fifth, and fourth [4].

The Pythagorean school thus regarded music as a science on a par with mathematics, and the study of proportions was a fundamental part of their teaching, with the aim of finding a numerical law underlying the world. The Pythagoreans thus discovered three types of mean: the *arithmetic, geometric,* and *harmonic means*.

Regarding the arithmetic mean, if two numbers a and b have the property $a < b$, there exists c such that

$$c - a = b - c \tag{3.1}$$

By a simple manipulation, we derive c, which is called the arithmetic mean and expressed as follows:

$$c = (a+b) : 2 \tag{3.2}$$

If we take the numbers 3 and 5, the mean will be

$$(3+5) : 2 = 4 \tag{3.3}$$

The geometric mean b of two numbers a and c is also called the mean proportion and satisfies

$$c : b = b : a \tag{3.4}$$

Taking a new triplet of numbers 2, 4, 8, we have

$$8 : 4 = 4 : 2 \tag{3.5}$$

The harmonic mean c of two numbers a and b has the following property:

$$(c-b) : (b-a) = c : a \tag{3.6}$$

For the triplet of numbers 6, 8, 12, we have

$$(12-8) : (8-6) = 12 : 6 \tag{3.7}$$

Curiously, the philosopher Archytas discusses these means in a text devoted to music, since the Pythagoreans regarded astronomy, geometry, arithmetic, and music as related sciences, meeting at a crossroads, the sciences of the quadrivium.

The quadrivium was conceived by Pythagoras in the form of the tetraktys, arising from his main interest in numbers and contrasted with the trivium, which was founded on the values of truth, beauty, and goodness and brought together grammar, logic, and rhetoric. The quadrivium promises an understanding the nature of the universe through the four associated sciences. Arithmetic consisted of three levels: the materially countable, indefinite, and archetypal or ideal numbers. Geometry comprised four levels: the point, which as it moves becomes a line, which in turn generates a plane, and finally creates a tetrahedron. Harmony was composed of four scales: pentatonic, diatonic, chromatic, and shruti. Astronomy dealt with the cosmos; the visible sky was ordered by pure principles, and the number of planets was related to harmonic proportions [5]. Before Pythagoras, other philosophers had tried to understand what kind of matter the universe was made of. Thales was convinced that it was water, while for Anaximenes it had to be air, Heraclitus entrusted the task to fire, and Anaximander argued that the primordial element had to be intangible and unmeasurable. Pythagoras did not believe that the universe consisted of tangible elements; he believed that the primordial substance was Number since *"everything is rational (number), which encodes belief in the mathematical intelligibility of nature"* [5]. Pythagoras had inherited calculating techniques from the Egyptians. Thales had already learned a method for calculating the height of a pyramid by measuring the shadow cast on the sand, and the Rhind Papyrus bore the title *Rules for Obtaining Knowledge of All Things Obscure*. The concept of cosmic music was certainly much older than the Classical Greek period; the god Thoth had linked planetary harmony with music, and in depictions from the Middle Kingdom, a harpist is seen playing a six-stringed harp, and on his head are painted six red discs without inscriptions [6]. A find from the Temple of Hathor in Dendera refers to the goddess as the holder of cosmic music:

The sky and the stars make music for you.
The sun and the moon pray for you.
The gods exalt you.
The goddesses sing for you. [7].

The discs found may suggest the existence of musical concepts. It therefore seems possible that the depictions are evidence of a Universal Harmony in the astronomical-musical sense. It would not be hard to believe that, after his studies in Egypt in 535 BCE, and upon his return to Italy, Pythagoras may have brought with him the beliefs and knowledge he imposed in the Pythagorean school. During his studies on musical acoustics, the mathematical physicist and physiologist Hermann von Helmoltz argued as follows:

"[…] if, as is probable, his (Pythagoras') knowledge was derived in part from Egyptian priests, it is impossible to know in what remote epoch this law was first known […]" [8]. This hypothesis is plausible when we consider that the Egyptians had already discovered the beauty of certain mathematical relationships, such as the Sacred Triangle [9] of Isis, Osiris, and Horus, mentioned by Plutarch and associated with the numbers 3, 4, and 5, the same numerical triad that would arise in Pythagoras' theorem for right-angled triangles.

For the Pythagoreans, the first four whole numbers formed the basis of their arithmetic and were represented in the *tetraktys* (see Fig. 3.1), a pyramid formed by these numbers. From the first four whole numbers, we find the main intervals of the whole of harmony, and their consonance is a function of the simplicity of the numerical ratios. From the tetraktys, we can derive the

3 Everything Is Number. Pythagoras and the Pythagoreans

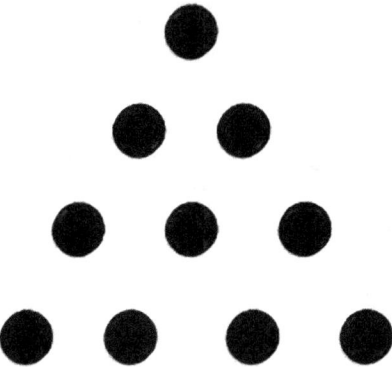

Fig. 3.1 The tetraktys

ratios 4:3 (fourth), 3:2 (fifth), 2:1 (octave), and unison. Furthermore, the sum of the first four numbers is 10, the number considered to represent the whole, or the universe.

Philolaus, a disciple of Pythagoras, wrote: "*[The number had become] perfect and omnipotent and the guiding principle of divine and heavenly life and of human life. [...] Without it all things would be unlimited and obscure and incomprehensible*" [9].

The cosmos of the Pythagoreans was schematized by Philolaus as a series of concentric spheres such that each planet was attached to a sphere rotating from west to east around a central fire located at the center of the Universe, with all the fixed stars attached on the outermost sphere. The Earth was such that its inhabited side was always facing in the opposite direction to the central fire, which explained why it was not possible for the inhabitants of the planet to see this central fire (see Fig. 3.2).

Philolaus argued that "*all things happen by necessity and harmony*" and the heavenly bodies, as they rotate, emit sounds in harmonic proportions, whence the universe produces a song based on rhythmic and harmonic motion [10]. Unfortunately, however, human beings inhabiting planet Earth cannot hear this song. The Pythagoreans explain this by arguing that Man is addicted to such singing, since he has heard it since birth, but Aristotle, in *De Caelo*, would object that sound is perceived if opposed to silence, and since we do not know the "sound" of silence, we cannot hear the song of the celestial spheres. Aristotle shared the cosmo.

logical model in the astronomical treatises of his predecessors and provided the following explanation of the relationship between the speed of the stars and harmony:

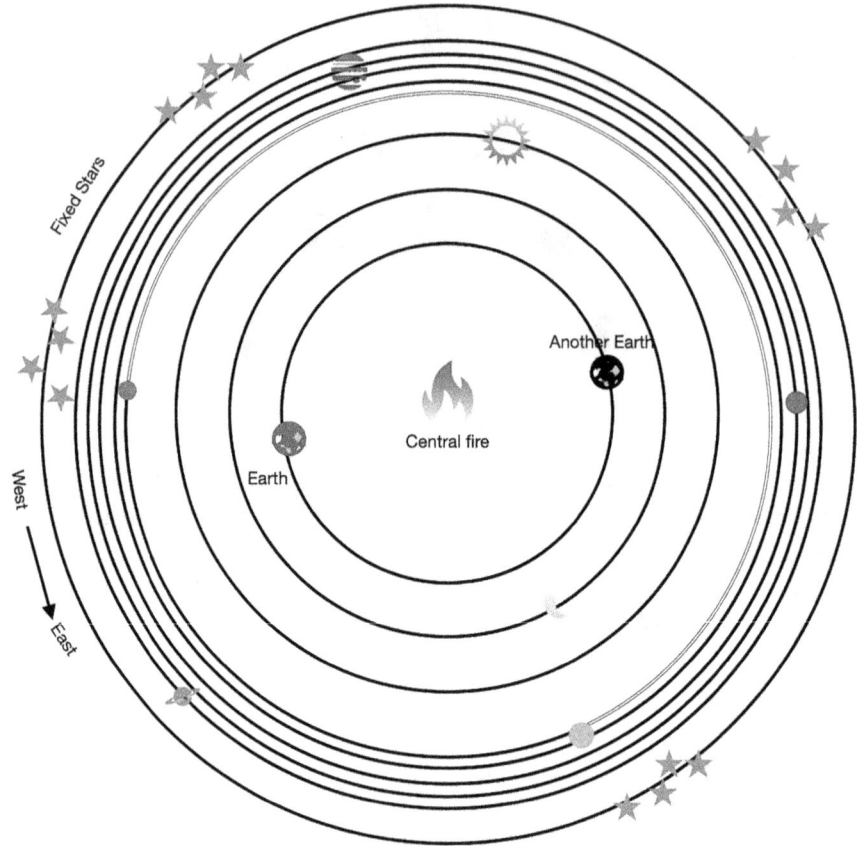

Fig. 3.2 The cosmos according to Philolaus

> *[…] harmony is generated from the motion of the stars; since the noises they produce would be consonant with each other, […] the velocities determined by the distances should be in the same proportions as the ratios of the consonances […]* [2].

Archytas gives another explanation for the impossibility of hearing such cosmic music: "Many of these noises cannot be perceived by our nature, some because of the weakness of their impact, others because of their great distance from us, some even because of the very excess of their intensity; because noises too great do not penetrate our ear […]" [2].

According to the philosopher, air particles, when they collide, transmit sound to the ear and brain. Archytas is thought to have been the first to suggest bringing together the disciplines of geometry, arithmetic, music, and astronomy in the quadrivium, and, as a follower of the Pythagorean doctrine,

3 Everything Is Number. Pythagoras and the Pythagoreans

he would have understood the relationship between the pitch of a sound and its speed of propagation, and he would also have understood why no interval can be divided into equal parts.

The problem of consonance brings the discussion back to the anecdote of the blacksmith's hammers. Using a monochord, a string attached at both ends and stretched above a sound box, Pythagoras was able to experiment with the relationship between the lengths of the strings and the intervals. The most significant discovery was that the harmonies that are considered pleasing are produced by the simplest mathematical proportions. When a string is plucked, a vibration is created that causes the string to oscillate up and down (see Fig. 3.3).

Pythagoras' experiment can be repeated using a string of unit length attached at the ends. If it is set in free vibration, the sound it produces will be proportional to the length of the string and is said to emit the fundamental note. By clamping the string to half its length, it will emit twice as many vibrations per unit time, generating the sound called an "octave." Blocking the string again at one-third of its length will produce a sound that is two-thirds of the fundamental note, called a "fifth."

Musical intervals can be constructed by multiplying the corresponding ratios together. So, to get the octave, 2:1, we can multiply a fifth by a fourth:

$$3:2 \times 4:3 = 2:1 \tag{3.8}$$

To solve Archytas' problem of dividing a musical interval into two equal parts, we must find the *mean proportion* of two numbers a and b. The solution would thus be

$$a:x = x:b \tag{3.9}$$

That is,

$$x = \sqrt{ab} \tag{3.10}$$

Since all musical intervals are of the type

$$\frac{n+1}{n} \tag{3.11}$$

Archytas proved mathematically that there is no mean value, so it is impossible to divide intervals into equal parts [11]. He referred to his Pythagorean

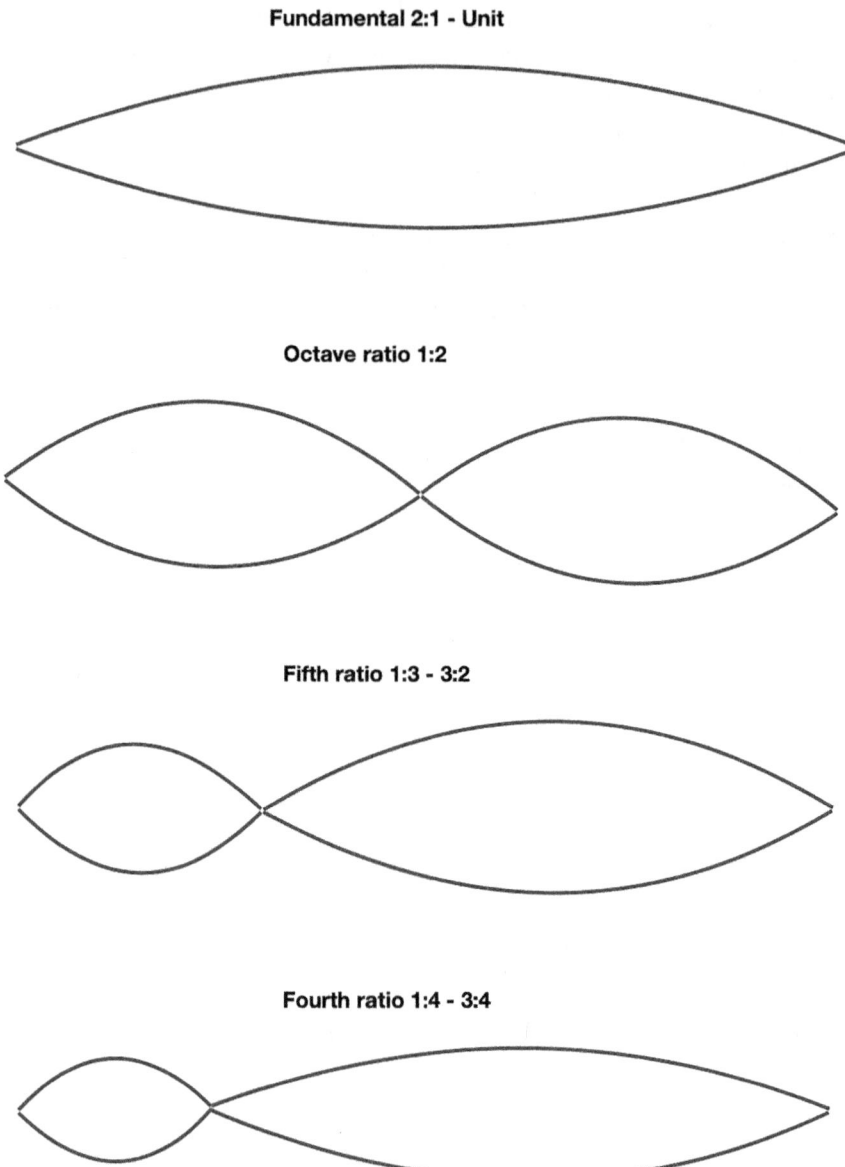

Fig. 3.3 Harmonic ratios

predecessors as *geometers* and *musicians* before they were arithmeticians, which suggests that celestial harmony was already practiced a generation before Archytas, that is, during the time of Philolaus [2].

The Pythagoreans assigned the Earth–Moon distance to unity, while the Earth–Sun distance stood in the ratio of 2:1, and Venus was at three times the

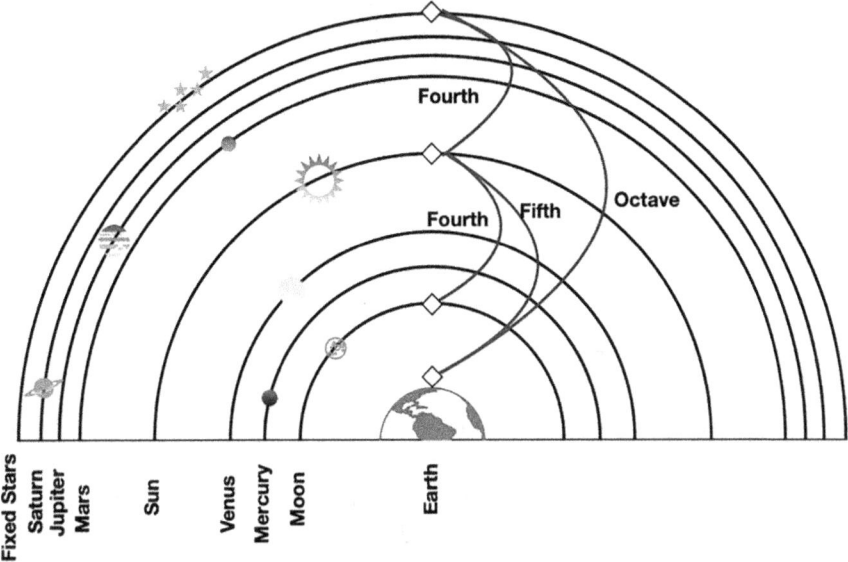

Fig. 3.4 The celestial spheres according to the Pythagoreans and as described by Censorinus

unit distance, hence in a ratio of 3:1 with regard to the Moon and 3:2 to the Sun. With this procedure, the sound emitted by each planet could be determined from the speeds of revolution and their distances. According to Censorinus, "*the world is a heptachord because it has seven planets which possess the bulk of the motion,*" and in his description of the cosmos he positions the celestial spheres according to harmonic ratios (see Fig. 3.4).

In this depiction, the Earth appears motionless in the center, since Aristotle replaced the idea of a central fire by a geocentric theory that would finally answer the question: "What does the Earth rest upon?" Aristotle states: "*The noises produced by them would be consonant with each other, […] and, assuming that the velocities determined by distances stand in the ratios of the consonances, those philosophers affirm that the rotation of the stars generates a harmonious sound*" [2]. Thus, the velocities of the planets were a function of their distances and established according to numerical ratios of consonances. If one planet moves at twice the speed of another, then the sound it produces will be an octave above the sound of the other planet.

Plato takes up the Pythagorean idea by stating that musical harmony is a reflection of the perfection existing in the world of numbers and is manifested in the motion of the planets. In *The Republic*, he uses the myth of Er to explain his planetary system with eight moving spheres, in which the speed of

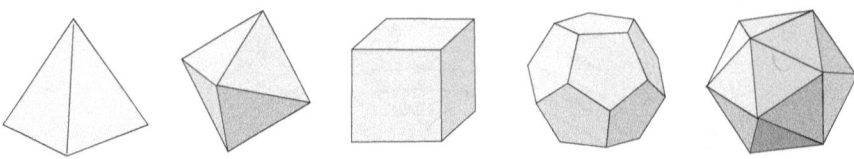

Fig. 3.5 The Platonic solids: tetrahedron, octahedron, cube, icosahedron, and dodecahedron. Public domain

the spheres depends on the distances, thus adopting order of the planets proposed by Philolaus. To each sphere he assigned a Siren which emitted a single note, always of the same pitch and intensity, and all together they formed a harmony.

If in *The Republic* Plato relies on myth, in *Timaeus* he seeks a more scientific explanation of the harmony of the spheres and describes the creation of the universe by a benevolent demiurge who imprinted a mathematical pattern upon the otherwise formless primordial matter [12]. The Platonic universe came into being with the formation of the five fundamental solids: the cube, octahedron, icosahedron, tetrahedron, and dodecahedron (see Fig. 3.5).

The first four solids were related to the natural elements identified by Empedocles: earth, air, water, and fire. The fifth solid, the dodecahedron, was associated with the aether, a crystalline solid from which the universe was made.

In Plato's cosmos, distances from the Earth can be evaluated with reference to two geometric progressions:

$$1,2,4,8$$
$$1,3,9,27$$

Plato assigned the unit distance from the Earth to the Moon in order to give the distances as follows: Moon 1, Sun 2, Mercury 3, Venus 4, Mars 8, Jupiter 9, Saturn–fixed stars 27. These are the orbital radii to each celestial object in terms of the Earth–Moon distance (see Fig. 3.6).

The philosopher did not consider the Earth when evaluating these distances because it was not itself considered to have any motion. And remaining motionless, it would not therefore generate any sound. Hence, the interplanetary spaces were seven in number. Interestingly, the first four powers of 2 and the first three powers of 3 occur, so we have the squares 4 and 9 and the cubes 8 and 27. By arranging this series in a triangular array, we obtain precisely the Pythagorean tetraktys (see Fig. 3.7). With these seven numbers, it is possible to construct the musical scale by simply multiplying or dividing by a certain

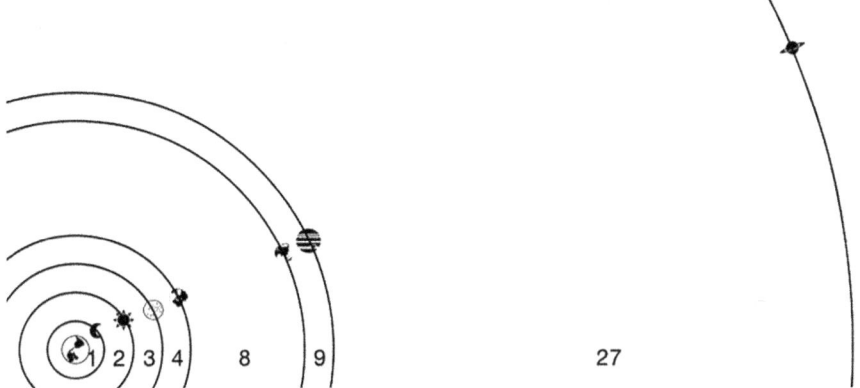

Fig. 3.6 Orbital radii in the Solar System according to Plato's theory

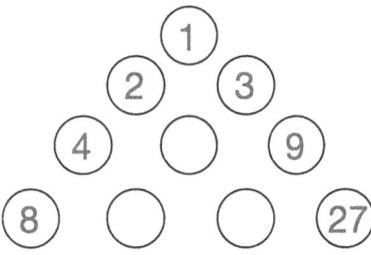

Fig. 3.7 The Pythagorean tetraktys superimposed on the number series of Plato's cosmos

ratio. In modern terms, the notes on the staff are represented on a logarithmic scale [5].

It is easy to see that, the notes from F to A are all ratios in which the numbers of the sequence mentioned above succeed each other, while E and B have quite unusual ratios (see Table 3.1). Clearly, dividing the ratios by 9/8 will yield the values 27, 8, and 4, showing that the sequence of seven numbers recurs within E and B.

The mathematics of the *Timaeus* is thus a "musical geometry." In the creation, matte,r took shape according to musical criteria [13], so we should not be surprised at Reinach's claim that "*it is not astronomy that has dictated its terms to music; on the contrary, it is musical practice that has left its mark on astronomical theories*" [14].

Music is thus arithmetized in the sense that it is attributed the creative power of number, which in the *Timaeus* is precisely the entity capable of creation. The harmony of the cosmos becomes the perfect model from which

Table 3.1 Correspondence between notes and numerical ratios

Tone	F	C	G	D	A	E	B
Ratio	2/3	1	3/2	9/4	27/8	81/64	243/32

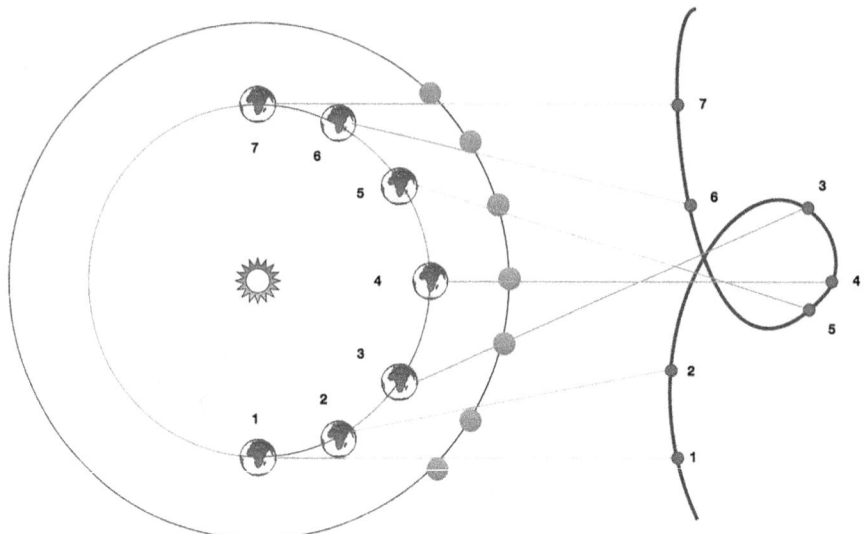

Fig. 3.8 Graphic representation of retrograde motion

musical harmony flows. Number is also responsible for the affinity between music and the human soul since Man is also part of Nature and his soul is sensitive to natural laws.

One of the most pressing problems of the fifth century BCE was to provide an explanation for the apparent retrograde motion of certain planets. Eudoxus of Cnidus hypothesized that the motion was the result of a celestial mechanism that involved at least three spheres rotating on different axes [15]. In the second century AD, Claudius Ptolemy formulated a series of mathematical models for each of the planets then known and succeeded in reproducing the observed apparent retrograde and periodic motion of the planets (see Fig. 3.8).

The model in use at that time was of the epicyclic–deferent type, devised by Apollonius of Perga (second cent. BCE), in which a planet P moves around an epicyclic circle whose center, for its part, moves around a deferent circle centered on the Earth. When the planet passes by points A and B, an observer on Earth sees it moving in the opposite direction to its motion relative to the fixed stars. In the *Almagest*, Ptolemy constructed an equivalent model with a moving eccentric, in which planet P moves around a circle centered on an

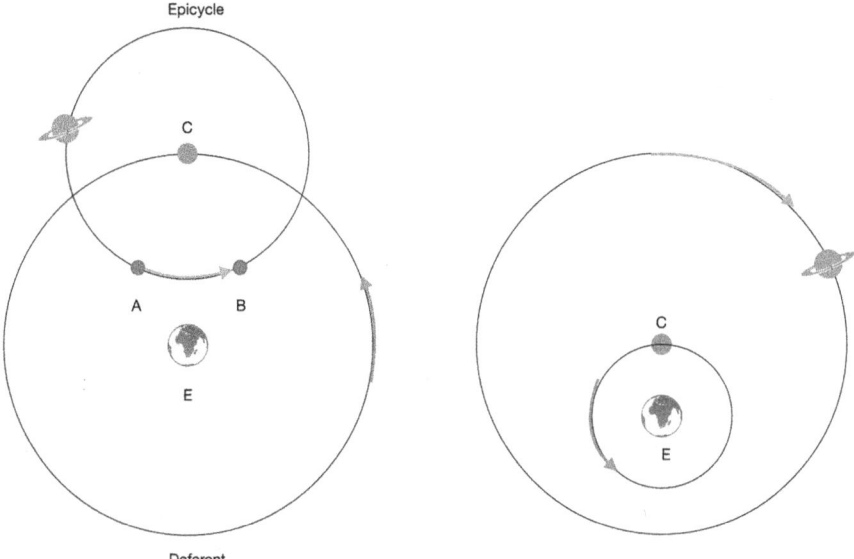

Fig. 3.9 Epicycle–deferent model and moving eccentric

eccentric point C that moves in the opposite direction around a circle with the Earth E as its center (see Fig. 3.9).

Ptolemy also devoted himself to the study of music. He believed that *"the instruments of judgment in harmonic science are hearing and reason, [...] hearing with regard to matter and accidental condition, reason with regard to form and cause"* [16].

A novelty in his model of harmonic science is the relationship between consonant intervals, the main features of the soul, and the longitudinal motions of the stars, in practice combining music, ethics, and astronomy. For Ptolemy, the explanation of consonance lay in the relationship with the soul of man and with the entire cosmos. He clearly based his understanding on the Pythagoreans, founding his harmonic science on the metaphysics of Plato and Aristotle, and on Greek cosmology.

References

1. C.L. Joost-Gaugier, *Measuring Heaven* (Cornell University, Ithaca, 2006)
2. A. Barbone, *Musica e filosofia nel Pitagorasmo* (La scuola di Pitagora editrice, Naples, 2009)

3. M. Giangiulio, *Pitagora. Le opere e le testimonianza*, vol II (Mondadori, Milan, 2000)
4. M. Timpanaro Cardini, *Pitagorici. Testimonianze e frammenti*, vol 3 (La nuova Italia, Florence, 1958–1964)
5. P. Odifreddi, *Penna, Pennello e Bacchetta,le tre invidie del matematico* (Editori Laterza, Rome–Bari, 2005)
6. M. Agrò, *L'Antico Egitto e la musica* (Ananke, Turin, 2009)
7. H. Junker, Poesie der Saatzeit, in Zeitschrift für ägyptischenshe Sprache, (43, 190 g)
8. H. Helmoltz, *Sensations of Tone* (London, 1895)
9. S. Isacoff, Temperament., *How Music Became a Battleground for the Great Minds of Western Civilization* (Vintage Books Edition, Random House Inc., New York, 2003)
10. A. Frova, *Armonia Celeste e dodecafonia, musica e scienza attraverso i secoli* (BUR, Milan, 2006)
11. G. Valerio, Genesi ed evoluzione della matematica (self-published)
12. J. Losee, *A Historical Introduction to the Philosophy of Science* (Oxford University Press Inc., New York, 1972)
13. N. Di Stefano, *Consonanza e dissonanza, teoria armonica e percezione musicale* (Carocci Editore, Rome, 2016)
14. T. Reinach, La musique des sphères, in Revue des études grecques 13 (1900)
15. P. Oliva, *Cosmogonie e Cosmologie* (La Lepre Edizioni, Rome, 2018)
16. C. Tolomeo, La scienza armonica, in *La scienza armonica di Claudio Tolomeo. Saggio critico, traduzione e commento*, ed. by M. Raffa, (Sfameni, Milan, 2002)

4

The Music of the Spheres in Kepler's Cosmos

The fervent Pythagorean Nicholas Copernicus (1473–1543) was convinced that mathematical harmonies really existed in physical phenomena. In *De revolutionibus*, he observed that his heliocentric system was able to explain the frequency of occurrence of the retrograde motions of the planets and was thus able to reveal the true harmony of the universe, in contrast to the Ptolemaic system, which could not provide any explanation whatsoever for these events.

Johannes Kepler (1571–1630) believed that the universe was created by a geometer God. His cosmos consisted of six planets and five regular solids. In *Mysterium Cosmographicum* (1596), he showed that the distances of the planets, considered as lying on concentric spherical shells, could be related to the five Platonic solids (see Fig. 4.1).

Kepler had no financial support from his family. His studies were paid for by a scholarship and he made extra money by doing horoscopes. Astrology enabled him to enjoy the good will of Graz's high-ranking families, but he was well aware that this kind of prediction was based on pure nonsense.

In Kepler's time, scientific research was still habitually interwoven with mystical interpretations, so some historians tend to consider him and Tycho Brahe as transitional figures in astronomy, placing them between Copernicus and Galileo. At the University of Tübingen, mathematics was taught by Michael Maestlin, a teacher who, in the lecture theatre, explained the Ptolemaic system approved by the Reformed Church, but in private took it upon himself to introduce the new Copernican system to a reserved group of students, including Kepler. Because he had contracted smallpox at a young age, Kepler could not expect to achieve any noteworthy observational results,

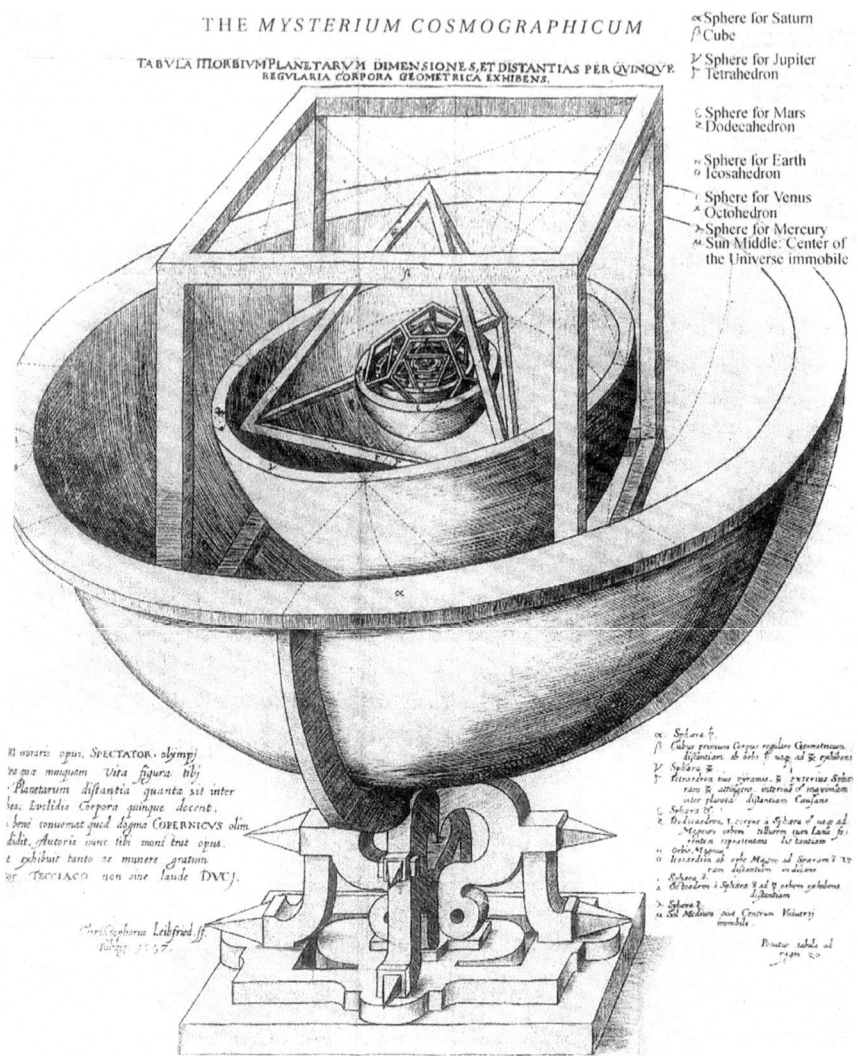

Fig. 4.1 Kepler's model. Johannes Kepler, *Mysterium Cosmographicum*. Public domain

and for this reason, he relied on the method of the ancients; in other words, he used reason to seek an explanation for the nature of the cosmos.

A first task was to explain why there were exactly six planets, according to the Copernican model, considering the Earth as a planet. Enlightenment came one day when trying to explain the conjunction of Saturn and Jupiter in the triangle of the Zodiac formed by the constellations of Aries, Leo, and Scorpio, a conjunction which occurs every eight hundred years. On the blackboard, he drew a triangle inscribed in a circle, and with another circle inscribed within it, as shown in Fig. 4.2 [1]. The triangles represented the major conjunctions of Jupiter and Saturn. Kepler the scientist understood that the relationship between the two circles must be the same as the relationship between the orbits of the two planets and the center of the universe.

This led to the idea that the distances should be related in some way to the regular solids enunciated by Plato. Placing the solids inside concentric spheres resulted in six spheres, one for each planet. The model was based on the mystical belief that the cosmos was governed by geometry.

In the early 1600s, Kepler moved to Denmark to work with astronomer Tycho Brahe. Unfortunately, Tycho died the following year, leaving behind all his observations, characterized as they were by the accuracy of his data. Kepler

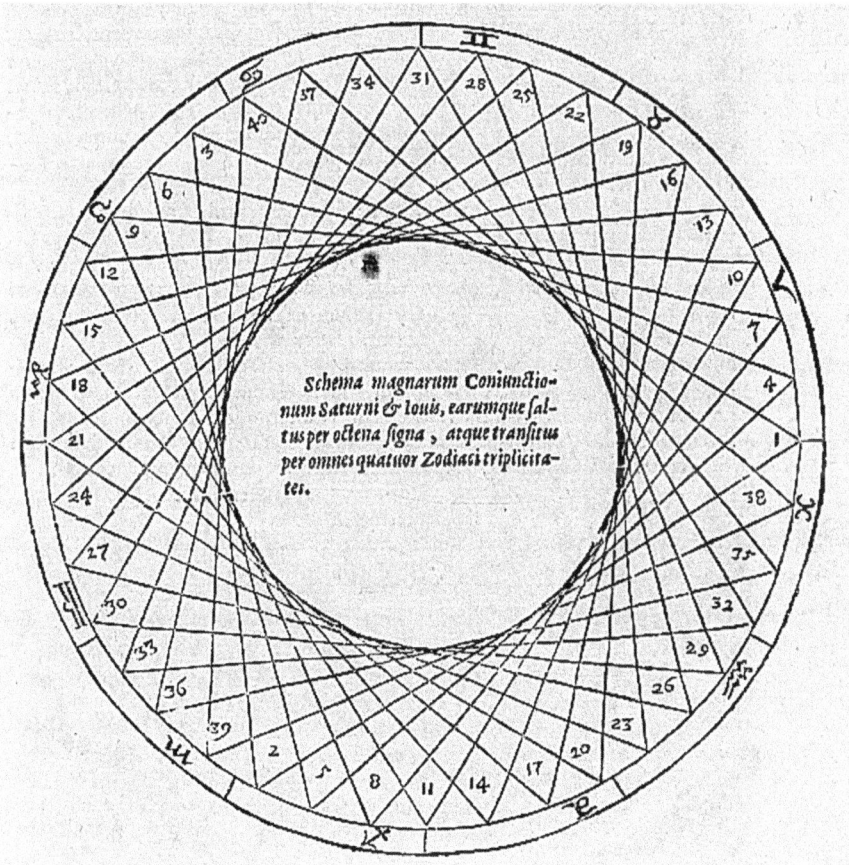

Fig. 4.2 Diagram showing the major conjunctions between Jupiter and Saturn. Johannes Kepler, *Mysterium Cosmographicum*. Public domain

rearranged and published them in the Rudolphine Tables (1627). However, he found that the geometric model of the cosmos did not match the data from Brahe's observations. His model then fell apart because it turned out that the planets did not move on circular orbits. The most noted problem was given by the orbit of Mars. Brahe's data showed that it was not after all a circle. When Kepler represented the results graphically, he found that the orbit was actually elliptical. When he calculated the orbits of the other planets, he found that they all followed the pattern of elliptical orbits, and thus stated his first law: *the planets move in elliptical orbits and the Sun occupies one of the foci* (see Fig. 4.3).

The idea that the Sun occupied one of the foci was like a poetic reference to the "central fire" or "hearth of the universe" so dear to the Pythagoreans [2].

During his research, he noticed that Mars moved faster when it was in the half of its orbit closest to the Sun and more slowly in the other half. He then imagined a line joining the Sun to the planet as it moved on its orbit and formulated his second law: *a line between the Sun and the planet sweeps out equal areas in equal times* (see Fig. 4.4).

These first two laws form part of what Kepler himself called the new astronomy, published in *Astronomia nova* in 1609 after a considerable delay due to lack of funds for printing. The book was not successful among scientists of the day because they were not yet ready to accept the heliocentric theory, nor the idea of elliptical orbits. Official recognition of his findings came only when Isaac Newton combined his laws with universal gravitation to explain the motion of planets on elliptical orbits.

One of Kepler's last works was his *Harmonices Mundi* (1619), in which he expounded his famous third law, also called the *harmonic law*. Going back to Kepler's belief that God was a musician as well as a geometer, it would have been perfectly unthinkable for the planets to emit the same note all the time, since the world would have been monotonous and musically unremarkable [3]. In *Harmonices Mundi*, he quotes a comment by Proclus on the first book of Euclid's *Elements*, in which Proclus defends the Platonic doctrine that states

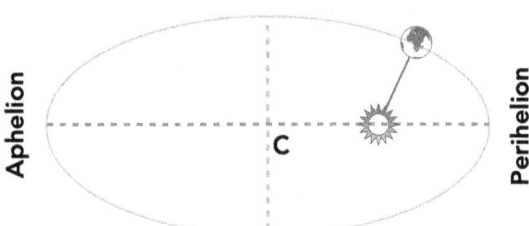

Fig. 4.3 Representation of Kepler's first law

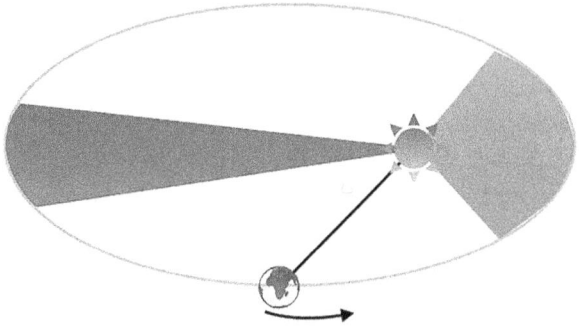

Fig. 4.4 Representation of Kepler's second law

Table 4.1 Relationship between geometric figures and musical intervals

Ratio of one side to the whole			Residual	
Diameter	1/2:1	Octave	1/2:1	
Triangle	1:3	Twelfth	2:3	Fifth
Square	1:4	Double octave	3:4	Fourth
Pentagon	1:5	Double octave + major third	4:5	Major third

that mathematical ideas are innate in the soul, while Aristotle considered them to be "universals abstracted from manifold sensory experience" [4]. Kepler thus rejected the numerological view cherished by Pythagoras, as well as the Platonic numerology expressed in the *Republic*, replacing it with a geometric view on the grounds that the cause of harmonies is not numbers as abstracted from sensory objects, but numbers as descriptors of real objects and thus geometric forms associated with the motions of celestial bodies. Kepler thus imagined a musical structure in Copernicus' heliocentric system, thereby justifying the idea of a "harmony of the spheres," although there was nothing scientific in his reasoning. In *Mysterium Cosmographicum*, he had guessed that musical harmonies and the five regular solids arise from the five plane figures. In fact, using only a ruler and compass, it is possible to divide the circumference of a circle into equal parts to construct the diameter, an equilateral triangle, a square, and a pentagon [5].

In truth, he had got this idea from Ptolemy's *Harmonics*, in which Ptolemy related the twelve intervals that make up the double octave to the twelve signs of the zodiac [1]. The arcs of circles obtained in this way allow certain mathematical relationships that can be traced back to the fundamental consonances (see Table 4.1) [1].

So, if *"everything is music"* [2], then the relationship between the times of revolution and the mean radii of the orbits must also be "music"–this was the reasoning that led Kepler to the formulation of his third law. Studying the relationship between the length of orbit travelled per unit time at aphelion and perihelion, he became convinced that there was a correspondence with musical intervals, and therefore constructed a celestial symphony by assigning a voice and a timbre to each planet (see Table 4.2) [6].

It thus seemed that the universe created by God was in reality governed by the rules of the musical scale. In *Harmonices Mundi*, Kepler described how the idea of the third law finally came to him:

[…] the actual relation between the periods and the dimensions of the planetary orbits […] was slowly conceived on March 8 of this year sixteen hundred and eighteen, but unfortunately, when calculations were made, it had to be rejected as false; but in the end, taken up again on May 15 with renewed ardour, light dispelled the darkness from my mind and confirmed that sixteen years of my labours on Brahe's observations and the considerations of this work of mine had finally come together in a single point, in such a way that at first I thought I was dreaming and that I had assumed the result of my investigations among the principles [used to obtain them]. But it is absolutely certain and exact that the ratio between the periods of any two planets is exactly the sesquilateral ratio of their average distances, that is, of the orbits themselves. [1]

The ratio discovered had to lie somewhere between the linear and the quadratic, which in music translate to the unison and the octave. This ratio corresponds to the fifth 3:2, and these numbers are precisely the exponents of the mathematical formalization of the third law, which can be summarized as follows: *the square of the period of revolution is proportional to the cube of the mean radius of the orbit:*

$$\frac{a^3}{T^2} = \text{const.}$$

Table 4.2 Voice and musical interval for each planet

Planet	Ratio	Interval	Voice
Mercury	12:5	Octave+minor third	Soprano
Mars	3:2	Fifth	Tenor
Saturn	5:4	Third	Bass
Jupiter	6:5	Minor third	Bass
Earth	16:15	Semitone	Alto
Venus	24:25	Pythagorean comma	Alto

4 The Music of the Spheres in Kepler's Cosmos

To paraphrase Odifreddi, the astronomer discovered the third law by deriving it from "patently nonsensical" but "miraculously correct" considerations [7]. An important implication of this law is its application to the relationship between the Sun and the other planets.

Going back to Copernicus'observation of the differing motions of a given planet along its orbit according to its distance from the Sun, he had formulated the hypothesis that the planets were kept in motion on their orbits by some kind of force, emanating from the Sun, which he called *vis harmonica*. This "force" was weaker at greater distances from the Sun and so propelled the planets further away more slowly.

The ratio between the force of attraction of the Sun and the square of the distance of any given planet is constant. In practical terms, Kepler had laid the foundation for the law of universal gravitation formulated by Newton.

Book V of *Harmonices Mundi* is devoted to the *music of the spheres*. Kepler assigned each moving planet a musical note based on his third law, thus giving rise to a true musical score (see Fig. 4.5). Each planet runs through a fixed interval of the musical scale, and it is notable that the Earth corresponds to the diatonic semitone 16:15, i.e., the interval E-F, commented on by Kepler with the exclamation "Oh miserable planet, so musically flat!" The astronomer goes on to give a rather mystical justification for this interval: "*The Earth sings E, F, E: just from these syllables, you can deduce that there is nothing but misery and hunger in this world*" [8].

Fig. 4.5 Excerpt from Book V of Kepler's *Harmonices Mundi*. Public domain

Kepler's music of the spheres was not monodic; he viewed the music of the world as a polyphonic motet, modelled on Orlando di Lasso's motet *In me transierunt* [9]. This polyphonic "song" was obviously a theoretical construction and not music that was actually played. Kepler himself deduced that the celestial symphony was played only once at the act of Creation and would be played again on the Day of Judgment [6].

References

1. G. Uggias, Keplero e la musica. Il Libro III dell'Harmonice Mundi (Linz 1619), Traduzione e introduzione, (PhD Thesis, Università di Bologna, Ciclo XXVII, 2015)
2. P. Odifreddi, *Dalla Terra alle Lune* (Rizzoli, New York, 2017)
3. A. Frova, *Armonia celeste e dodecafonia* (BUR, Milan, 2006)
4. D.P. Walker, La musica celestiale di Keplero, in *P. Gozza, La musica nella Rivoluzione Scientifica del Seicento*, (Il Mulino, Bologna, 1989)
5. J. James, *The Music of the Spheres. Music, Science and the Natural Order of the Universe* (Grove Press, NewYork, 1993)
6. P. Odifreddi, *Penna, Pennello e Bacchetta, le tre invidie del matematico* (Editori Laterza, Rome–Bari, 2005)
7. P. Odifreddi, Dalla Terra alle Lune (Rizzoli, 2017)
8. Keplero, Harmonices Mundi, Cap. V
9. P. Pesic, *Music and the Making of Modern Science* (MIT Press, Cambridge, Massachusetts, 2014)

5

Scientist Musicians

Before Kepler, the keeper of the Pythagorean tradition was Severinus Boethius (480–524) author of the treatise *De instituzione musica*, in which he divided musical expression into three areas: *Musica Humana, Mundana*, and *Instrumentalis*. His model was Platonic: sound was produced by an impulse and percussion, and these could not occur without motion, so sound was defined as a continuous percussion of the air right up to the point of hearing [1]. For Boethius, theoretical treatises were superior to actual musical practice in the sense that *"it is far more urgent and important to know what one does, than to implement what one knows; for material artifice serves as a slave, while reason commands almost as a lady [...] How superior is the science of music in theoretical knowledge, compared to practical implementation!"* [2].

In his tripartition, *Musica Instrumentalis* was the part dealing with "practical implementation," i.e., the actual music, to which he gave a negative connotation, illustrating the Greek view of manual work as not being worthy of the free man [3]. *Musica Humana* belongs to the soul of man and reflects the music of the spheres. Finally, *Musica Mundana* is precisely the celestial music of the spheres. For Boethius, this was indeed the true music: *"[It] is discernible especially in those things which are observed in heaven, or in the combination of elements, or in the diversity of seasons. Is it possible that so swift an organism should move with such tacit and silent motion?"* [4].

The fact that it was impossible to hear the music of the cosmos presented no problem to Boethius. Indeed, he took this very impossibility as an indication of its perfection. The "sound" was thus to be identified with the concept of harmony, in conformity with the ideas of the Platonic and Pythagorean tradition. Boethius played the role of ferryman between the theoretical treatises of antiquity and medieval philosophical thought. He brought the ideas of the Hellenic world to his own contemporaries, filtered through Pythagorean thought. As a result, it is not difficult to trace the doctrines of the Pythagorean school down through the Middle Ages. The first to write a treatise that would reshape the disciplines of the quadrivium and the trivium was Alcuin, a minister of Charlemagne. He placed music alongside the sciences in a pyramidal scheme:

Philosophy
Ethics, physics, logic
Arithmetic, music, geometry, astronomy, astrology, mechanics, medicine

Music was defined as the "discipline dealing with the numbers to be found in sounds" [4]. All medieval thought was underpinned by the idea that there is a mathematical musical principle, but these doctrines founded on the thought of Boethius and adapted by Augustine would gradually fade with the approach of the Renaissance. Johannes Tinctoris was one of the first music theorists to break with medieval tradition and move away from mathematical explanations of music. He dealt only with *Musica Instrumentalis*, defining harmony as "something pleasant produced by appropriate sounds."

The existence of a close link between music and astronomy or mathematics is demonstrated by the many musicians who mastered the quadrivium arts over the centuries, as illustrated by the English composer and astronomer John Dunstable who lived during the 1400s. Even earlier, between 1151 and 1158, the Benedictine nun, composer, and naturalist, Hildegard of Bingen, wrote the work *Symphonia armonie celestium revelationum*, a collection of texts set to music, showing that music is the highest expression of praise to the harmonies of Creation as represented by the heavenly spheres. For Hildegard, music evoked the celestial consonance that reigned in Paradise before the original sin [5]. In this sense, music was both celestial and terrestrial, since only with *Musica Instrumentalis* would it be possible to reproduce the music of the cosmos. We have to wait for Gioseffo Zarlino and other scientist musicians before we see a return to *Musica Mundana*.

In his three major works *Istituzioni harmoniche* (1558), *Dimostrazioni harmoniche* (1571), and *Sopplimenti musicali* (1588), the Venetian Zarlino (1517–1590), a pupil of the composer Adrian Willaert, sought to refound musical theory on the natural foundations of sounds. Hence, the concept of *Musica Mundana* would form the background supporting the idea that the ratios between the musical intervals are not arbitrary, but rather that there is a rational relationship based on the "nature of things." So, the music of the celestial spheres is a recurring theme which, however, can no longer be traced back to its purely theoretical and inaudible connotation, but should rather be understood as a function of the rationalization and mathematization of music. The *Istituzioni harmoniche* is a truly encyclopedic treatise, ranging over topics from harmony to history, mathematics, and even musical theory and practice. In this work, he cites all the most important exponents of the classical period, from Pythagoras to Plato, Aristoxenus, Ptolemy, Augustine, and Boethius.

According to Zarlino, *"all things created by God were ordered by Him with numbers: indeed number was the chief exemplar in the mind of the maker"* [6]. It is no coincidence that his search for the perfect number went back to the Pythagorean tradition, and the *Senario* became the cornerstone of his theory. The number 6 takes on the value that the tetraktys had for the Pythagoreans: there were six days in the Creation; six is the first perfect number since it is equal to the sum of its factors 1, 2, and 3; it is also equal to the product of its factors; and finally, there are six planets in the Solar System [7]. Excluding the decomposition into its factors, the other two reasons are rather fanciful. The number 6 also encompasses all the consonances obtainable from the first six numbers, and in fact had the consonances of the third and sixth intervals, which the Pythagoreans had considered to be dissonant, were rehabilitated in the Renaissance [8].

Hence, Zarlino identified the "universal reason" in numbers, and believed that music could restore the balance between body and soul and put the parts of the world back together, all this being possible because it was a system based on harmonic proportions. However, the system put forward in the *Senario* posed a problem, because the ratio that gives rise to the minor sixth is 8:5, and the number 8 does not fit into the associated series of numbers. Furthermore, if he had made 8 fit in with this reasoning, he would also have had to accept the number 7. Zarlino dismissed the problem by arguing that the number 8 is implicit in the series, since it is either twice 4 or the cube of 2, and in this case, we return to the Pythagorean numbers that form the interval of the fifth.

This solution was not appreciated by his pupil Vincenzo Galilei (1520–1591), lute player, theorist, scientist, and father of the famous Galileo Galilei. In his treatise *Dialogo sulla musica antica et della musica moderna* (1581), Vincenzo discussed the formal theoretical principles of modern music and compared them to Greek music, regretting the lost role of ancient music, especially with regard to its educational function. His other major dissatisfaction was the Pythagorean origin of Zarlino's theory. Galilei considered it unacceptable to demonstrate consonance through the use of the tetraktys or *Senario*, which expressed the need to find a valid tuning system based on a scientific model. In his view, the arithmetic component arose from a mental exercise that ignored an important peculiarity of music, namely that it modified the soul and provoked feelings [1].

The Renaissance saw a move away from the mathematization of music and its consideration as a science, toward a greater interest in its artistic and emotional aspects, as exemplified by the emotions stirred when listening and the use of the spoken word. It is not therefore surprising that Vincenzo Galilei would extol the "emotional and cathartic virtues of singing" [9] and proclaim the supremacy of early music due to its ability to elevate the spoken word. In his *Sylva Sylvarum*, Francis Bacon also supported the idea that consonance could not just be a question of numbers. He was convinced that the field of investigation had to be extended to the perception of and reaction to sound stimuli, thus anticipating to some degree the work of Hermann Helmoltz in the nineteenth century.

The new style of music flourished at the *Camerata de' Bardi* in Florence, where musicians and poets gathered in the house of Giovanni Bardi for the purpose of re-founding music and use of the spoken word. The dispute with Zarlino became quite bitter. The maestro, and now rival, accused Vincenzo of musical relativism, an assault on God's plan. But before the publication of his *Discorso intorno alle opere di Gioseffo Zarlino* in 1589, Galilei made a fundamental discovery. For a long time, theorists had believed that Pythagoras had used the same proportions to determine the weights applied to the strings, that is, applying different weights to the same lengths of string, resulting in the ratios 2:1, 3:2, and 4:3. Galilei realized that when weights were applied to the strings to obtain the same result, they had to be in the ratios 4:1, 9:4, and 16:9, i.e., the weights were not in inverse proportion to the length of the string, but in inverse proportion to the square of its length [10].

The position of Vincenzo Galilei was much closer to that of Giordano Bruno, who admonished the poets in his remark: "*Throw off the yoke of authority; there are no rules other than the ones you make*" [7]. In his work *La cena de le ceneri* (The Ash Wednesday Supper), Bruno embraced Copernicus' heliocentric theory, but went beyond his finite universe by considering it to have no center and no boundaries:

> *Whoever says there are infinitely many worlds, they must all be included in a single space and the same sky must contain all things. Your contrivance consisting of spheres and stars and which revolves around the Earth with diurnal motion does not correspond to the world [...] for me the Earth is a world, similar to the Moon, and far away, and not dissimilar, is the Sun and all the bodies that sparkle in the great space, adorning the stages of Olympus with admirable order* [11].

Bruno thus advances a strong criticism of the Aristotelian model of the cosmos, in which *"there can be no more worlds than one"* [12], thereby freeing himself intellectually from pre-Copernican theories and attempting to go beyond the boundaries of the Solar System by the "force of reason" alone. For the philosopher from Nola, the universe was an "infinite sphere," with *"as many centers of the world as there are stars"* [11]; and again, *"there are therefore innumerable Suns, and there are infinitely many Earths, which circle those Suns just as we see these seven circle our own Sun"* [13].

When he was summoned by the Church and refused to retract his claims against geocentrism and the existence of other worlds, he declared: *"You were more afraid in passing this sentence than I was in receiving it"* [14]. He was subsequently burned at the stake on February 8, 1600.

The existence of other worlds had been theorized much earlier by Epicurus: *"there is an infinity of worlds, some like ours, some different, and there is no reason to believe that in these other worlds there are no species of animals or plants, and everything else we see here"* [15]; and the idea was taken up later by Lucretius, so Giordano Bruno was just unlucky to live in the wrong era.

However, this was the historical period in which Vincenzo Galilei was active, a period of intellectual revolution in which his son Galileo, reluctant to accept the official doctrine, encountered the same resistance as Bruno. Music had gone through the Protestant Reformation and the Counter-Reformation of the Church, so was not immune to attacks against new ideas, either regarding its practical conception in the form of liturgical chant, or its theoretical basis. Bruno's infinite universe, in open contrast to Zarlino's, thus translated into Vincenzo Galilei's musical universe of infinite numbers, while Galilei's son Galileo used his father's discoveries to formulate a theory based on science. The scientific method attributed to Galileo was derived directly from the rules for a "new music" established by Vincenzo Galilei, Giulio Caccini, and the Florentine Camerata.

The model was based on the sequence of ideation, development, contradiction, affirmation, elaboration, and presentation. These are basically the same rules as the scientific method, but surprisingly, they are also the rules that would give birth in Viennese classicism to the sonata form, consisting of exposition, development, and recapitulation. The new theories of consonance can be traced back to Giovanni Battista Benedetti, who in 1585 was the first to understand the physical particuliarities of sounds and was probably the main inspiration for Vincenzo and Galileo Galilei. The music of the spheres began to lose its appeal, and mystics like Zarlino and Kepler were superseded by mechanists who began to give concrete meaning to the numerological criteria that had dominated up until the Renaissance [16]. Modern acoustical science, based on observation and calculation, was born with the arrival on the scene of seventeenth century scientists like Marin Mersenne (1588–1648), René Descartes (1596–1650), Christiaan Huygens (1629–1695), and Jean D'Alembert (1717–1783). Moreover, music would break away from the disciplines of the quadrivium as the emerging *theory of affects* gradually replaced Pythagorean and Platonic values with a desire to express and codify feelings at the expense of mathematical considerations. Music would thus be perceived as *the art of forging sounds*, as Stravinsky put it, and its connection with the cosmos and the celestial spheres, in the ancient sense, would gradually be forgotten.

References

1. N. Di Stefano, *Consonanza e Dissonanza, teoria armonica e percezione musicale* (Carocci Editore, Rome, 2016)
2. S. Boezio, Pensieri sulla musica, ed. by A. Damerini, (Fossi Editore, Florence, 1949)
3. E. Fubini, *Estetica della Musica* (Il Mulino, Bologna, 1995)
4. E. Fubini, L'estetica musicale dall'antichità al '700, (Piccola Biblioteca Einaudi, Milan, 1976–2002)
5. M. Gargantini, Santa Ildegarda di Bingen profetessa teutonica (in Insegnare religione, Nov. 1998, ElleDICI, Turin)
6. G. Zarlino, Istituzioni Harmoniche, Venezia, 1558, libro I, cap. XII (ed. By Ridgewood, Gregg, 1966)
7. S. Isacoff, *Temperament. How Music Became a Battleground for the Great Minds of Western Civilization* (Vintage Books Edition, Random Hose, Inc, New York, 2003)
8. M. Agrò, Storia dell'austica: un connubio tra matematica e musica (in Rassegna Musicale Curci, ed. Curci Editore, Milan, LXIII no. 2 May 2010)
9. G. Gallico, *L'età dell'Umanesimo e del Rinascimento* (EDT, Turin, 1991)
10. D.P. Walker, in *Vincenzo Galilei e Zarlino*, ed. by P. Gozza, (La musica nella Rivoluzione Scientifica del Seicento, Il Mulino, Bologna, 1989)
11. G. Giorello, E. Sindoni, *Un mondo di mondi* (Raffaello Cortina Editore, Milan, 2016)
12. Aristotele, De caelo, I 8, 276 b15
13. G. Bruno, De l'infinito, universo e mondi
14. M. Frigerio, *Invito al pensiero di Giordano Bruno* (Mursia, Milan, 1991)
15. Epicurus, Letter to Herodotus, in Diogenes Laertius, Lives of Eminent Philosophers
16. A. Frova, *Armonia celeste e dodecafonia* (BUR, Milan, 2006)

6

From Newton to Einstein

During the seventeenth century, music and astronomy seemed to take different paths, with the former following a path related to art and the latter beginning its process of mathematization. However, the concept of the music of the spheres would continue to link the two disciplines, but in an entirely new way, and the debate would increasingly appeal to the tools of mathematics.

The naturalist philosopher Gottfried Wilhelm Leibniz (1646–1716) would state that "*music is a hidden arithmetic exercise of a mind that does not realize that it is manipulating numbers*" [1]. This idea went decidedly counter to the ideas of Bacon and Galileo, and showed that the Pythagorean conception of music lived on. In his letters to Maurice Solovine, Einstein wrote: "*mathematics is but a means of expressing the laws that govern phenomena*" [1], a thought not dissimilar to the ideas expressed by René Descartes (1596–1650), who considered musical science to be objective and its laws derived from nature. Music was thus based upon simple and predetermined mathematical formulas. It is no accident that the philosopher rejected the equal temperament system, because it does not start from the simplest musical proportions, but adapts them to how pleasant they sound to the ear. Descartes considered music to be a perfect candidate for his project of unifying mathematics and physics using arithmetic and geometry, and in *Compendium musicae* he analyzed rhythm and consonance from a mathematical point of view.

Isaac Newton (1642–1727) reconsidered the Pythagorean perspective on the music of the celestial spheres and traces the discovery of the law of universal gravitation back to Pythagoras' experiments with strings and weights. In his most important work, the *Principia*, he inserted some scholia in Propositions IV–IX of Book III – known as classical scholia because some of them refer to ancient Greek and Roman wisdom – in which we find some remarks on music. According to Newton, "*[...] I thought Pythagoras's music of the spheres was intended to typify gravity and as he makes the sounds and notes to depend on the size of the strings, so gravity depends on the density of matter*" [2]. The unusual attribution of gravitation to Pythagoras was a way of tying in with the Pythagorean and Platonic tradition by showing that his work was merely a rediscovery of ancient wisdom through mathematics [3]. The outline of Scholium VIII explicitly refers to the experiment in which Pythagoras found the law of universal gravitation as the inverse ratio of the square of the lengths of the strings:

$$t = \frac{1}{l^2} \tag{6.1}$$

The tensions in the strings thus turn out to be related to the square of distance. Newton wrote:

> *By what proportion gravity decreases by receding from the Planets the ancients have not sufficiently explained. Yet they appear to have adumbrated it by the harmony of the celestial spheres, designating the Sun and the remaining six planets, Mercury, Venus, Earth, Mars, Jupiter, Saturn, by means of Apollo with the Lyre of seven strings, and measuring the intervals of the spheres by the intervals of the tones. [...] For Pythagoras [...] stretched the intestines of sheep or the sinews of oxen by attaching various weights, [...] and consequently by comparing those weights with the weights of the Planets and the lengths of the strings with the distances of the Planets, he understood by means of the harmony of the heavens that the weights of the Planets towards the Sun were reciprocally as the squares of their distances from the Sun. But the Philosophers loved so to mitigate their mystical discourses that in the presence of the vulgar they foolishly propounded vulgar matters for the sake of ridicule, and hid the truth beneath discourses of this kind. [...] Pythagoras beneath parables of this sort was hiding his own system and the true harmony of the heavens*" [2]. In fact, Newton knew that Pythagoras got the ratios wrong, as evidenced by Newton's work *Of Musick*, which was never published and remained in handwritten form. For the first time, he put forward analogies between music and optics that would be taken up again in his book *Opticks*, where he compared the spectrum of colors with the seven notes of the diatonic scale [4]. For Newton, music, light, and the planets were all part of the same divine harmony. Indeed, he argued that the laws governing gravity must be the same as those governing the behavior of musical strings. At this time, the idea of a new temperament was in the air, a system that could codify tuning in a universal way. Newton thus studied various temperaments and even proposed some tunings, but Newton's

musical work remained only a marginal contribution to the temperament problem [5].

The scientific method and advances in the mathematical formulation of physical phenomena changed the way the problem of consonance was interpreted from a mystical conception related to the need to understand the cosmos to a scientific conception that emphasized the psychological and physiological significance of consonance. The Age of Enlightenment provided a basis for the emergence of a musical aesthetic that contemplated consonance as a source of enjoyment for the ear, and in the Classical–Romantic era any mathematical reference in music was eschewed in favor of the expression of feelings. Descartes was convinced that *"the purpose of music is to delight and inspire different feelings in us"* [6], while for Leibniz mathematical structure is oblivious to music and the enjoyment experienced in consonances was only one of its manifestations.

The nineteenth century philosopher Eduard Hanslick had mixed views for his time. In his book *On the Musically Beautiful*, he explained his reasons for claiming that music expresses nothing and is unsuitable for arousing emotions. Hanslick's discussion is connoted by a strong scientific objectivity, for he insists that the investigation of what is beautiful, if it is not to become completely illusory, must inevitably adopt the methods of the natural sciences [7]. So, for Hanslick, music is devoid of purpose and the ideas expressed by the musician are first and foremost purely musical. As for Helmoltz, music can represent the dynamics of feelings, but not express them. According to Hanslick, *"mathematics merely controls the intellectual manipulation of the primary elements of music, and is secretly at work in the most simple relations. The musical thought, however, originates without the aid of mathematics [...]"* [8].

The second half of the 19th century was characterized by the "scientification" of music studies, thanks to the drive of positivism and Hanslick's thinking. Psychological, psychophysical, acoustic, and anthropological investigations became a major part of the attempt to answer Carl Dahlhaus's question: "What is music?" While Helmoltz went back to the work of Zarlino and Rameau to lay the foundations of musical psychophysiology, Darwin, Spencer, and Combarieu were the first to study the origins of music. In the *Descent of Man* (1877), Charles Darwin wrote:

> As neither the enjoyment nor the capacity of producing musical notes are faculties of the least use to man in reference to his daily habits of life, they must be ranked amongst the most mysterious with which he is endowed. They are present, though in a very rude condition, in men of all races, even the most savage [9].

The concept of *Musica Mundana* was gradually abandoned in favor of what Plato hated most, namely *Musica Humana* and *Instrumentalis*. In fact, modern acoustic science no longer looks at numbers in connection with the universe, but focuses on the mathematical modeling of sound and its direct effects on the human psyche. However, the connection with the cosmos remains one of the essential prerogatives that can be found, no longer in direct philosophical thought, but in instrumental work, and even more so in the life and work of physicists.

Max Planck (1858–1947), the father of quantum mechanics, began studying piano as a child. In his high school days, he did not stand out in science subjects, but was rather gifted as a performer, devoting himself to the works of Brahms, Schubert, and Bach (see Fig. 6.1).

Planck could definitely have had a career as a professional pianist, but he was dissuaded by a musician friend and so "fell back" on physics and mathematics [10]. Planck was also a composer of songs and small pieces, not to mention a chamber opera. He played Beethoven with the violinist Joseph Joachim, and he performed in a quartet with his son Erwin, a cellist, Albert Einstein, and the physicist Arnold Sommerfeld. As a student, Planck attended Hermann von Helmoltz's physics courses at Berlin's Friedrich Wilhelm University, and this brought him indirectly into contact with Hanslick's thinking on mathematics and music. It was Sommerfeld who eventually returned to the ancient theme of the music of the universe. Speaking of his work on the analysis of spectra he stated that "*[...] the language of spectra is a*

Fig. 6.1 Max Planck at the piano. Public domain

true 'music of the spheres' internal to the atom, a symphony of admirable relations, an order and harmony that becomes ever more perfect [...]" [11].

Among the musician scientists, it is also worth mentioning Hugh Chistopher Longuet-Higgins (1923–2004), who studied chemistry and music at Oxford and made important contributions to quantum chemistry, apart from earning four honorary degrees in chemistry and even a degree in music [12]. His contribution to music is condensed in a number of publications on the perception of rhythm and sensation. Werner Karl Heisenberg, known for his discovery of the *uncertainty principle*, also studied piano, reaching a level suitable for him to give public performances. During one of his concerts, he met Elisabeth Schumacher, who became his wife in 1937.

In astrophysics, the Stefan–Boltzmann constant (the constant σ in the formula σT^4 for the energy radiated by a black body) plays a decisive role in determining the brightness of stars, and few people know that Ludwig Boltzmann, small in stature and stocky, had studied piano with the well-known composer Anton Bruckner, author of nine hefty symphonies [13].

The physicist who made the most important contribution to research on the cosmos was Albert Einstein. As a child, he studied the violin. He had his own violin which he called Lina, and which he kept with him throughout his life. Together with his friend Planck, he enjoyed chamber music, performing Bach, Mozart, and Beethoven. Critics gave contradictory judgments, but Shinichi Suzuki considered him to have a "pleasantly delicate" timbre, while others considered his play rough compared with the musical elegance of Max Planck. For Kurt Gödel, there were many musicians with much better technique, but he felt there was no one who touched the strings of a violin more sincerely or more deeply than Albert Einstein [14].

Several of his acquaintances reported that he "got a lot more excited over musical disputes than scientific disputes" [15]. During his studies he met Mileva Marić, a promising physicist but also a music lover, who played the piano and the tamburica, a type of lute native to the Balkans. Einstein visited her in the guesthouse where she lodged and asked her to play the Sonata No. 6 from Beethoven's Op. 30 No. 1 with him. He talked a lot about music with her and wrote pages of letters in which he revealed his passion for Bach. In one of these, he mentioned the Kyrie from Bach's Mass in B minor [14].

By the end of World War II, Einstein was no longer able to play the violin due to a hand injury, but the Juilliard String Quartet, visiting the now 70-year-old physicist's home, performed music by Beethoven and Bèla Bartok, and at one point, invited him to play with them. Einstein chose the second violin part of Mozart's Quintet in G minor [16]. In 1952 he became vice-president of the Princeton Symphony Orchestra.

Einstein's connection with music was fundamental to his research in physics. In an interview in 1929, he stated:

> "*If I had not been a physicist, I should probably have been a musician. I often think in music. My daydreams are filled with music, and I see my life in terms of music. I don't know if I would ever have created anything new or important in music, but I definitely know that playing the violin fills me with joy*" [17]. Music and physics were complementary because they stemmed from the same source, and music was able to catalyze his creative processes [14].

Music was not a mere diversion in Einstein's life. When he rented a house with his second wife Elsa, the couple decided to celebrate by giving a private concert with music by Haydn and Mozart. For the occasion, the great violinist Toscha Seidel played with Einstein, along with Arthur Giskin on cello and Bernard Ocko on viola (see Fig. 6.2). The artistic partnership with Seidel was based on a free exchange of information; in exchange for lessons on violin technique, Einstein attempted to explain relativity to the maestro [18].

Among the composers he could not stand was Wagner, of whom he said "*most of the time, I hate to hear him*" [14]. We do not know whether this

Fig. 6.2 Arthur Giskin cello, Toscha Seidel violin, Albert Einstein violin, Bernard Ocko viola in the house in Princeton. At Einstein's shoulder, his wife Elsa. Figure credit: @Leo Back Institute, F5333G

remark referred to the composer's musical style or his being anti-Semitic, since Wagner could well have reminded him of the Nazi period.

The father of relativity was a contemporary of Arnold Schoenberg, the father of twelve-tone music. For the composer Pierre Boulez, with the dodecaphonic system, music left Newton's world and entered Einstein's [19], and indeed dodecaphony played the same role in music as relativity in physics. In the early 1900s, unknown to most, Einstein was developing the theory of an unbounded universe and his principle of relativity, just as Schoenberg and the Futurists were abandoning tonality and Kandinsky was turning his hand to abstract art. The modern world was about to be born [20]. However, Einstein thought Schoenberg and his music were madness. The physicist and the musician met twice in 1934, first during a lecture by the musician at Princeton and then at Carnegie Hall in New York for a charity event (see Fig. 6.3).

Einstein's criticism of Schoenberg was not an isolated incident; the physicist had already expressed his opinion of several past composers. Of Handel, he said that "*Handel is good, even perfect, but he has a certain shallowness,*" and he did not spare Schubert either: "*Schubert is one of my favorites because of his superlative ability to express emotion and his enormous powers of melodic invention, but in his larger works I am disturbed by a certain lack of architectural shape.*" The search for structure inspired the following reflection on Debussy's

Fig. 6.3 Leopold Godowsky, pianist and composer, with Albert Einstein and Arnold Schoenberg. Figure credit: @Leo Back Institute, F5333H

music: "*Debussy is delicately colorful but [also] shows a poverty of structure.*" And regarding Schumann: "*Schumann is attractive to me in his smaller works, because of their originality and richness of feeling, but his lack of formal greatness prevents my full enjoyment.*" Only Bach and Mozart succeeded in grasping "*unity on a vast scale,*" and Mozart's work was "*so pure that it seemed to have been ever present in the universe, waiting to be discovered by the master*" [21].

Einstein's sensitivity to music led him, during one of his stays in Florence, to give his sister Maja (Maria Einstein) an 1899 Julius Blüthner grand piano acquired second-hand [22].

The relationship between physics and music remained close throughout the 20th century. Sir James Jeans, father of the theory of gravitational collapse, was a fine organist, and in 1938 published *Science and Music* [23]. Only eight years earlier, in *The Mysterious Universe* (1930), he stated that "*To my mind, the laws which nature obeys are less suggestive of those which a machine obeys in its motion than of those which a musician obeys in writing a fugue*" [24].

Regarding the auditory imperfections of harmonics, Jeans explains that "*[…] they arise out of the laws of arithmetic, which the musician is completely powerless to alter. If we visited another planet, we should find the same laws there as on earth*" [23]. Speaking of the music of the future, he added: "*As the laws of arithmetic would be the same on this planet as on earth, we conjectured that the inhabitants might quite possibly have arrived at the same musical scale as our own, the octave being divided into twelve equal, or approximately equal, divisions*" [23].

A particularly interesting feature of Jeans' book *Science and Music* is the presence on the first page of two Sumerian engravings, one from the royal tomb in Ur and the other from the tomb of Queen Shubad, depicting musicians and dancers. Jeans repeatedly cites ancient peoples to explain the physical phenomena of music, and does so with much reverence, as if sensing an ancestral link between music and the mathematical relationships that govern the universe.

The childhood of Wernher von Braun, father of rocket science, was marked by a love of music, and in particular composition. According to his mother, he could have had a career as a musician, as the young Wernher had taken cello lessons from Hermann Lietz at the Ettersburg Castel school near Weimar and joined the school's youth orchestra. He also occasionally played in quartets with Rudolf Hermann and Heinrich Ramm on violins and Gerhard Reisig on viola, performing Mozart, Haydn, and Schubert [25]. Already at the age of fifteen, Braun composed several works of a certain level that were strongly reminiscent of the compositional style of his piano teacher, the composer Paul Hindemith [26]. Nevertheless, the young Wernher considered himself a "frustrated pianist," when one evening in 1929 at Willy Ley's house he performed Beethoven's "Moonlight Sonata."

Physicists of the 20th century considered music to be closely connected to science, and not only from a purely mathematical point of view, but above all from a philosophical point of view. Music could help one to think (Einstein) or to explain the laws of nature (Jeans), or even to explain string theory. So, the music of the spheres had not completely disappeared, but would take on a new guise. Science would prove that there really was, and would continue to be, a cosmic music.

References

1. M. Livio, La sezione aurea. Storia di un numero e di un mistero che dura da tremila anni (BUR saggi, Milan 2003)
2. J.E. McGuire, P.M. Rattansi, Newton e la Siringa di Pan, ed. By Gozza P., La musica nella rivoluzione scientifica del Seicento (Il Mulino, Milan, 1995)
3. P. Odifreddi, *Penna, Pennello e Bacchetta, le tre invidie del matematico* (Editori Laterza, Rome–Bari, 2005)
4. P. Pesic, *Music and the Making of Modern Science* (MIT Press, Cambridge, Massachusetts, 2014)
5. S. Isacoff, Temperament., *How Music Became a Battleground for the Great Minds of Western Civilization* (Vintage Books Edition, Random House, Inc, New York, 2003)
6. N. Di Stefano, *Consonanza e dissonanza, teoria armonica e percezione musicale* (Carocci Editore, Rome, 2016)
7. E. Fubini, *Estetica della Musica* (Il Mulino, Bologna, 1995)
8. E. Hanslick, The Beautiful in Music, translated by G. Cohen, ed. by M. Weitz (Bobbs-Merrill Educational Publishing, Indianapolis 1957)
9. C. Darwin, *The Descent of Man* (Penguin, London, 2004)
10. L. Belloni, S. Olivares, *Planck, la rivoluzione quantistica*, vol 3 (Corriere della Sera, RCS, 2016)
11. A. Sommerfeld, Atomic Structure and Spectral Lines (Munich, 1919, translated by H.L. Brose, Dutton, 1923)
12. S. Califano, V. Schettino, *La nascita della meccanica quantistica* (Florence University Press, Florence, 2018)
13. M. Kumar, *Quantum: Einstein, Bohr and the Great Debate about the Nature of Reality* (Icon Books Ltd., UK, 2008)
14. G. Greison, *Einstein forever* (Bollati Boringhieri, Turin, 2020)
15. G. Farmelo, *It Must Be Beautiful. The Great Equations of Modern Science* (Granta Books, London, 2002)

16. C.V. Vishveshware, *Einstein's Enigma or Black Holes in My Bubble Bath* (Copernicus Berlin, Heidelberg, 2006)
17. V. Palermo, La versione di Albert: Perché Einstein è un genio (Hoepli, 2015)
18. W. Isaacson, *Einstein, his life and universe* (Simon & Schuster, New York, 2007)
19. E. Maor, *Music by the Numbers* (From Pitagora to Schoenberg (Princeton University Press, Princeton, New Jersey, 2018)
20. P. Griffiths, *Encyclopedia of Twentieth Century Music* (Thames and Hudson, London, 1989)
21. D. Bodanis, *Einstein's greatest mistake: a biography, Hogthon* (Mifflin Harcourt, Boston New York, 2016)
22. F. Palla, Il Pianoforte di Einstein, in Coelum Astronomia, no. 201 (2016)
23. J. Jeans, Science and Music (Dover Edition, 1968)
24. A. Balbi, *The Music of the Big Bang* (Springer, Heidelberg, 2008)
25. M. Wright, Wernher von Braun's support for the imaginary arts (NASA, NP-1999-06-78-MSFC)
26. R. Spangenburg, D.K. Moser, *Wernher von Braun* (Chelsea House, New York, 1995)

7

The Music of the Cosmos

During May 1964, two American radioastronomers, Arno Penzias and Robert Wilson, were working with a 6-meter radio antenna owned by Bell Laboratories in Holmdel, New Jersey (see Fig. 7.1). Their project was to adapt the receiver by fitting a Dicke radiometer for satellite communications and astronomical observation.

While they were trying to figure out what the internal noise level of the receiver was, so that they could subtract it from the observation data, they realized that they were picking up a very weak signal in all directions. After several attempts to understand the reason for this "hum," they decided to inspect the antenna, rolled up their sleeves, and began to remove "a white dielectric material," guano left by pigeons that had got into the antenna. The result was the same, however; the humming noise was still there in the background.

Several years earlier, Ralph Alpher and Robert Herman had shown that the formation of atomic nuclei in the primordial universe would imply the presence of a large amount of radiation, and that this radiation would remain as a residue in the present universe. The first theories suggesting the existence of a primordial radiation background date back to the period between 1948 and 1950, when several papers authored by Alpher, Herman, and George Gamow investigated the temperature these background photons would be expected to have in the present universe. Gamow had put forward a specific value of 3 K, as measured on the kelvin scale.

Returning to Penzias and Wilson's work, a few miles away was the Princeton dream team consisting of Robert Dicke, James Peebles, Peter Roll, and David Wilkinson. In 1965, Penzias had not yet solved the problem with the antenna,

Fig. 7.1 Penzias and Wilson's antenna. Figure credit: NASA

but while talking with Bernard Burke, a researcher at MIT, he learned that Peebles had given a seminar on the blackbody radiation that was supposed to pervade the universe. So, Penzias phoned Dicke, who, after hanging up, turned to his Princeton group and exclaimed, *"Boys we've been scooped!"*. The group went to talk to Penzias and Wilson and were soon convinced that the background noise was the primordial cosmic radiation at 3.5 K, a slightly higher temperature than Gamow had hypothesized. In 1978, Penzias and Wilson were awarded the Nobel Prize in Physics "for their discovery of the cosmic microwave background radiation" [1].

Prior to Einstein, the scientific concept of the cosmos had been firmly based on Copernicus' heliocentric model, Kepler's laws, Galileo's observations, and the theory put forward by Newton, who regarded the universe as eternal, since in his view space and time were absolute. The question of the nature of the Universe would be a constant in twentieth century physics and cosmology, and the Big Bang model was eventually accepted by the entire scientific community as a plausible explanation for the birth of our cosmos. This model for the evolution of our universe was born from Einstein's equations of relativity together with Edwin Hubble's observations of the receding galaxies and Georges Lemaître's and Alexander Friedmann's solution to the cosmological equation.

There had originally been another cosmological model called the steady state theory, proposed by Fred Hoyle, in which the universe continually produced new matter to compensate for its expansion, thereby remaining the same everytwhere at every instant of time. In contrast, the new model proposed by Lemaître, together with Alpher and Gamow, argued that the universe was expanding and cooling. The term "Big Bang" was coined ironically by Hoyle himself, who was its main detractor, during a BBC radio broadcast in 1949. When Penzias and Wilson discovered the cosmic background radiation, the Big Bang model finally took precedence over Hoyle's theory: the Universe was expanding from an initial singularity and the fossil radiation was evidence for its birth.

In the beginning, the universe was very simple, consisting of a plasma of atomic nuclei and free electrons interacting with a large amount of electromagnetic radiation, i.e., photons. This mixture was initially at a very high temperature, which meant that the the photons possessed high enough energies to engage in permanent interaction with the free electrons, bouncing around without being able to propagate freely. The phenomenon of radiation scattering made the universe completely opaque, as when a thick fog prevents light from passing straight through it, creating a soft light effect in the environment. The gradual cooling of the universe decreased the photon energies below the threshold that could prevent the formation of hydrogen atoms. This period is called recombination. At the same time, radiation and matter decoupled, and light was finally able to propagate throughout the cosmos: *"Then God said, 'Let there be light,' and there was light. And God saw that the light was good. Then he separated the light from the darkness"* (Genesis 1:3–25).

The photons freed at the time of decoupling are still travelling through the universe, and Earth-based antennas can detect this very weak radiation. From the cosmological point of view, the primordial plasma during recombination had a temperature of about 3000 K, but due to cooling, it looks today like black body radiation at about 3 K. The wavelength of the primordial photons was 0.001 mm, but the expansion of the universe led to cooling, in such a way that, each time its size doubled, the temperature halved, and in the present universe the photon wavelength has stretched to 1 mm, thus falling in the microwave range. For this reason, cosmologists refer to this fossil radiation as the *cosmic microwave background radiation.*

In 1989, NASA launched the Cosmic Background Explorer (COBE) satellite (see Fig. 7.2). It carried a very important instrument, the Differential Microwave Radiometer (DMR), which was a version of the Dicke radiometer used by Penzias and Wilson, designed to measure the temperature difference between pairs of different points. Another instrument, FIRAS, would

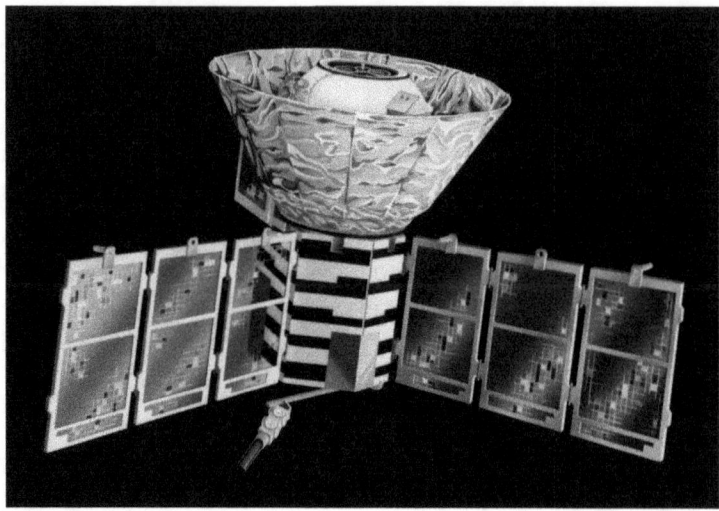

Fig. 7.2 COBE satellite. Figure credit: NASA

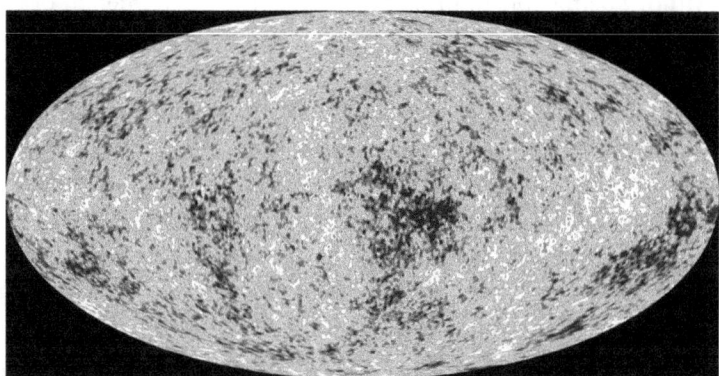

Fig. 7.3 Map of the cosmic microwave background. Figure credit: NASA

measure the energy distribution of the photons in the fossil radiation. By the end of the mission, it was possible to draw up a map of temperature fluctuations in the cosmic background radiation (see Fig. 7.3).

The cosmogonies of the ancient peoples now find a kind of scientific confirmation. Light, thunder, and the verb do indeed look like the seeds of the primordial universe in this microwave representation. Today's cosmologists have rediscovered the music of the spheres, the Musica Mundana dear to Boethius and Pythagoras, generator of the cosmos. The image produced by the ripples in the cosmic background radiation can be translated into an acoustic frequency spectrum by harmonic analysis, and this can then give us an idea of the intensity of the acoustic waves that must have passed through the primordial plasma.

As Brian Green declares, there is *"nothing but music"* [2], musical metaphors for meditating on the mysteries of the cosmos, and with superstring theory, music has once again come to play a prominent role in the search for the origins of our universe. The theory predicts that the point particles of the Standard Model of particle physics are made up of small, closed, vibrating strings. This idea goes back to the conclusions of Democritus and the atomists, because strings would be the fundamental constituents i.e., indivisible atoms in the Greek sense. Physicists refer to strings as if they were violin strings. Strings can vibrate in an infinite number of ways generating resonances that are perceived as musical notes. Strings have the same properties and different modes of vibration as a fundamental string, giving rise to various inertial masses and electric charges, just as a violin emits different musical notes with their various colors. So, the properties of a particle depend on the mode of vibration of its string, and one thus goes from a music of the spheres to a music of the strings.

The Pythagorean monochord model is once again central to the explanation of the formation and properties of particles. Just as the monochord can be plucked, a string can also be "plucked," not physically, but mathematically, and a "musical scale" of strings can be constructed in which each note represents a particular particle. In the words of Angelo Adamo, *"the idea that the strict mathematical rules underlying musical harmony are to be connected to the harmonies of universal progress is one that runs through the entire history of humankind, prompting some even to wonder if by any chance the universe might be nothing but a sound"* [3]. The universe would thus consist of a splendid musical score describing a cosmic symphony in which, according to Pope Benedict XVI, speaking of Beethoven's music, *"[...] human genius competes in creativity with nature, gives life to varied and manifold harmonies, where the human voice also takes part in this language, which is as a reflection of the great cosmic symphony"* [4].

References

1. A. Balbi, *The Music of the Big Bang* (Springer, Heidelberg, 2008)
2. B. Green, *The Elegant Universe* (W.W. Norton, USA, 1999)
3. A. Adamo, *Pianeti tra le note* (Springer-Verlag Italia, Milan, 2010)
4. X.V.I. Benedetto, *Sulla musica* (Marcianum Press, Venice, 2013)

8

The Universe Is Teeming with Aliens

The historian Eric Hobsbawm's "short century" began in 1914 with the advent of World War I. This conflict immediately presented itself in a completely new guise compared to the nineteenth-century wars, as it saw the deployment of new technologies and military strategies as well as the use, for the first time in human history, of aviation. This airborne revolution prompted US President William Howard Taft to approve the creation in 1912of a new body for aviation research. Thus, the National Advisory Committee for Aereonautics (NACA) was born. At the end of the two world wars, the Cold War became the main driving force, not only politically, but also and especially technologically. In 1957, the Soviets launched the first artificial satellite, Sputnik 1, into orbit. The Americans considered this launch a major threat, so in 1958, NACA was transformed into the new NASA (National Aeronautics and Space Administration) with the task of meeting the challenge in the space race between the two world superpowers, the US and the USSR.

It is natural to wonder what connection there could be between NASA, involved in space exploration, and music. Indeed, the two disciplines seem to be worlds apart: the former rational and scientific, the latter ephemeral and artistic. However, this is not the case. Music once again assumes a central role when viewed from a scientific and technological perspective.

In 1877, Thomas Alva Edison invented the phonograph, the first instrument capable of storing sounds and reproducing them, so it was inevitable that music would become its main application. The age of recording had begun. All the sounds of the world could be captured and stored, then listened to again anywhere and at any time. It was not until 1895 that the first 78-rpm records were first produced. This medium provided the impetus for the rapid development of the recording industry in the early 1900s. But records were not only important for music. They were also the tool of choice for NASA to send information about our planet and humanity into space.

The two space missions Voyager 1 and Voyager 2, launched on 5 September 1977 and 20 August 1977, respectively, carried with them a record of the existence of the human species contained in the *Voyager Golden Record* (see Fig. 8.1), a gold-plated copper record in which a message was recorded for the potential inhabitants of other worlds. Leading the committee that made the selection of what was to be recorded was the scientist and author Carl Sagan. The record contained the sounds of our planet, including wind, thunder, sea waves, animal vocalizations, a greeting from Earth translated into 55 languages, and a selection of music lasting 90 minutes.

Having made the record, there remained the problem of how to listen to it. For the inhabitants of planet Earth, the solution was within easy reach of anyone, of course, but this was not so for the possible inhabitants of other worlds. For this reason, instructions were included for using the record (see Fig. 8.2) along with a turntable stylus that could be used to build a suitable instrument for listening to it, somewhat like the instructions for building the interstellar travel machine in the film *Contact*, based on Sagan's own novel of the same name.

However, there was a second problem. Since the record carries with it a selection of music, this raises an obvious question: is music universal? If we consider the concept of music on the Pythagorean model, then we could say

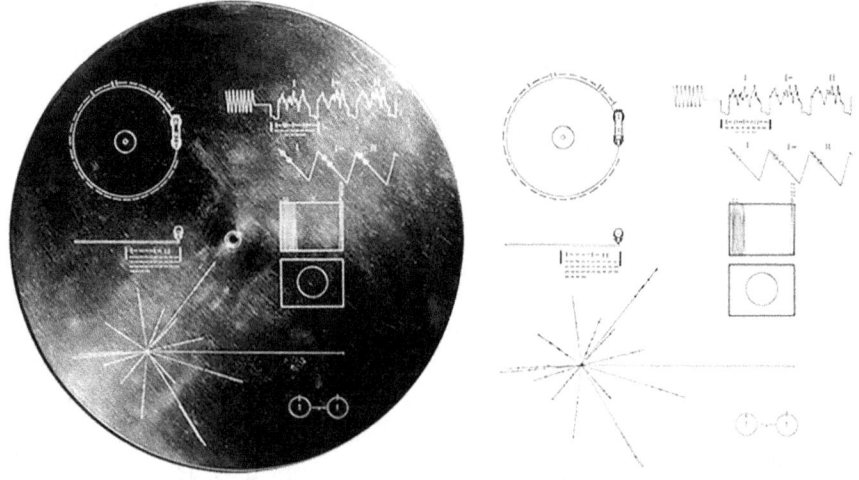

Fig. 8.1 Golden Record cover with instructions for reading the record. Figure credit: NASA/JPL

Fig. 8.2 Voyager Golden Record. Figure credit: NASA/JPL

that it is based on mathematical concepts, so an advanced civilization could easily decode the vibrations of the turntable stylus [1]. According to the creator of the project, Carl Sagan: "*The spacecraft will be encountered and the record played only if there are advanced spacefaring civilizations in interstellar space*" [2]. However, music is a human concept and not a natural one. An alien community might not have the same musical instinct as we do. In any case, the record represents an attempt to throw "a message in a bottle" into the sea of deep space, in the hope that someone might find it and understand it.

While the natural sounds collected around our planet and the information about who we are seem easy to justify, the idea of sending music is rather striking, whether we consider the concept of music in itself or its purely entertaining and scientifically non-functional nature. The launch of the two Voyager spacecraft was accompanied by the release of two cult films, George Lucas's *Star Wars* and Steven Spielberg's *Close Encounters of the Third Kind*. In this way, the year 1977 brought about new ways of imagining space, and it seems likely that it was this emotional impetus that gave Sagan the idea to communicate through music.

If an alien civilization ever encounters one of the Voyagers and listens to the record, it will have to consult an army of musicologists and musicians in order to interpret the meaning and value of what is recorded on it. The Golden Record aims to represent the music of the world, with typical songs from different peoples around the planet, and also classical compositions recognized the world over. So, the aliens will hear J. S. Bach with the first movement of the *Brandenburg Concerto No. 2 in F major*, the *Prelude and Fugue No. 1 in C major* from the "Well-Tempered Clavier," and the *Gavotte and Rondo* from the *Partita No. 3 in E major for violin*, but also Beethoven with the first movement of his *Symphony No. 5 in C minor* and the *Cavatina* from the *Quartet No. 13 Op. 130 in B-flat major*. Mozart also gets a place on the record, with the aria of the Queen of the Night taken from *The Magic Flute*, this being the only taste of opera included. The musical selection places the works of the great composers alongside Chuck Berry's song *Johnny B. Goode* and Louis Armstrong's *Melancholy Blues*, becoming the second Armstrong in space. In 1978, with some irony, a broadcast of *Saturday Night Live* on the American television channel NBC imagined the aliens' response to hearing the Golden Record: *"Send more Chuck Berry!"* [3].

While the idea of sending music into space on the Golden Record may raise eyebrows, on 4 February 4 2008, NASA beamed the Beatles' song *Across the Universe* directly into space to commemorate 40 years of the song and 50 years since the launch of Explorer 1, America's first artificial satellite. It was launched in the direction of the North Star Polaris, 431 light-years from Earth, upon which Paul McCartney commented: *"Amazing! Well done, NASA! Send my love to the aliens"* [4].

While on the one hand NASA is busy managing the music it sends out into space, on the other it is searching for and deciphering sounds coming the other way, from the depths of the cosmos. So, in the first case we can still speak of a Musica Istrumentalis, but in the second, we are dealing precisely with the Musica Mundana that was so dear to the Pythagoreans.

It was also in the summer of 1977, annus mirabilis for everything related to space, from the cinema to NASA, that Jerry Ehman, an astronomer at the Wesleyan University's Big Ear Radio Observatory, carefully analyzing the signals received by the antenna, came upon a particular sequence. The sheets on which the receiver data were printed took the form of numbers from 1 to 9 for lower intensities and letters of the alphabet as the intensity increased. The signal in question was encrypted as 6EQUJ5, and corresponds to a 72-second signal from the constellation of Sagittarius (see Fig. 8.3).

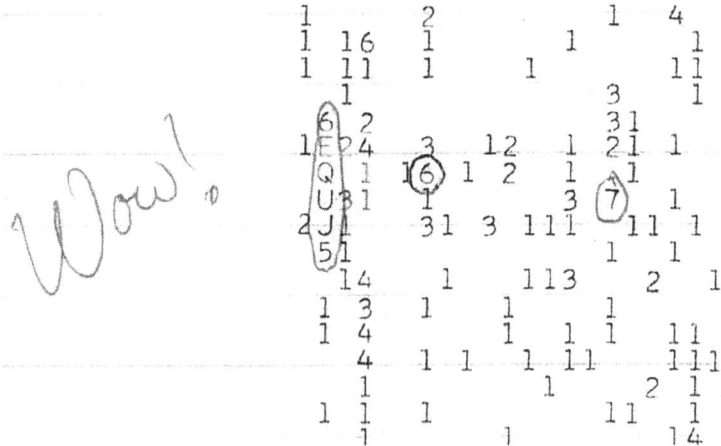

Fig. 8.3 The printout containing the Wow! Signal. Figure credit: Big Ear Radio Observatory and North American Astrophysical Observatory (NAAPO)

The peak intensity indicated by the letter U was at a frequency of 1420 MHz, a frequency that astronomers attributed to hydrogen in the 1950s. Ehman was so surprised that, on the printout of the data, he noted the word Wow! And from that moment on, this signal was called the Wow! signal. It was immediately thought that that string might be the result of an extraterrestrial intelligence that was communicating with us using the frequency of hydrogen, an element found in great abundance in the universe.

The idea that an alien civilization might be sending messages into space came from Ehman's own work as part of the SETI project, the search for extraterrestrial intelligence, whose main radiotelescope was located at Arecibo in Puerto Rico. However, the excitement ended in 2016 when, at the Center for Planetary Science in Florida, the signal was finally interpreted as being due to the passage of two comets and the hydrogen cloud surrounding them. While the Wow! signal is certainly not a message from another civilization, it is nevertheless one of the many voices of the universe that gives content to the concept of Musica Mundana.

On 14 January 2005, in collaboration with the Italian and European space agencies, NASA reached the surface of Titan, one of Saturn's moons. The Cassini–Huygens mission was launched in 1997 with the task of observing Saturn, its rings, and its natural satellites (see Fig. 8.4). The space probe consisted of an orbiter (Cassini) and a lander (Huygens) that was designed to enter Titan's atmosphere and descend right through it.

The first data on Titan go back to the measurements made by the Voyager probes during their rapid fly-bys. Huygens was thus equipped with

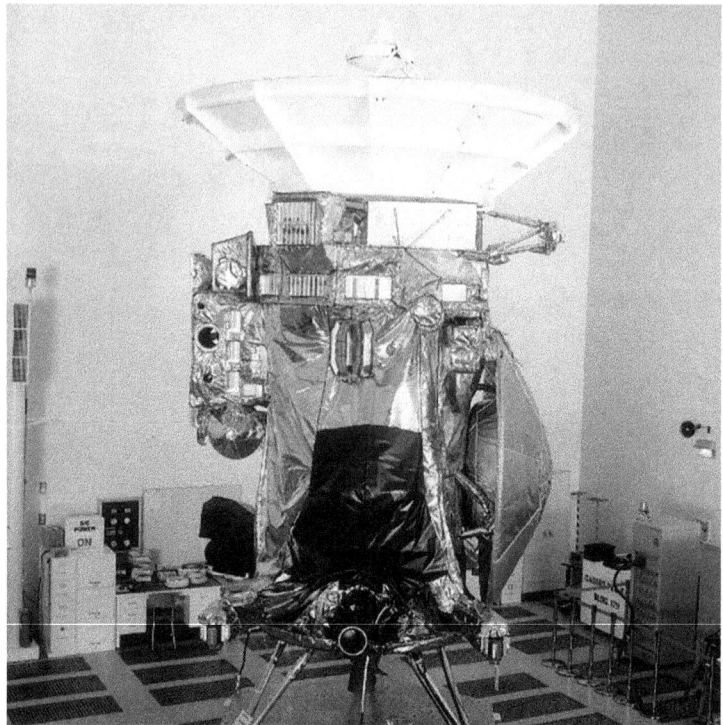

Fig. 8.4 Cassini–Huygens probe in NASA's laboratory. Figure credit: NASA

instruments to measure atmospheric parameters such as temperature, pressure, density, and electrical characteristics in greater detail. These instruments included a microphone to detect the presence of thunder, and hence lightning, in Titan's atmosphere, as this could trigger the chemical reactions needed to form organic compounds [5]. If the idea of Musica Mundana is based on the music emitted by planets during their orbital motion, then the sounds of Titan would surely be a good representation of such a manifestation.

Curiously, wandering around our Solar System, there are some remarkable sights to see. For one, there appears to be a disk playing in space: Saturn's rings (see Fig. 8.5) look just like the grooves of an old vinyl record (see Fig. 8.6), spinning on a cosmic record player, while the needle reading the sound is our probes, from Voyager to Cassini–Huygens. Looking at those images, we can admire the similarity in all its beauty. If the ancient Greeks could have observed images like this, taken from space, they would surely have taken Saturn for a cosmic DJ. The existence of a music of the universe would at once have become tangible.

8 The Universe Is Teeming with Aliens   63

Fig. 8.5 Saturn's rings. Figure credit: NASA

Fig. 8.6 The grooves of a vinyl record. Public Domain

When we observe the accretion disc around a protostar, we notice the same analogy with our vinyl records, although perhaps we should compare it to the more modern compact disc, given the density and brightness of the cloud. So, the rotation of cosmic objects and the existence of discs around such objects suggests that the universe really is a great musical machine.

The search for a music of the spheres reached its climax with the discovery of gravitational waves, ripples in space-time that propagate like sound waves, carrying the primordial cry, the creative voice that gave rise to the universe. Now we have further confirmation that the myths and legends about an initial creative sound have some foundation in truth.

Like sound waves, gravitational waves are also produced by a source that "vibrates," actually a large mass subject to violent acceleration. However, while sound waves are material oscillations, gravitational waves are oscillations of space. In 1916, Einstein published a paper claiming the existence of three different types of gravitational waves. Sir Arthur Eddington then proved that these waves were not real, but a construction based on the reference system used to make the calculations. In 1936, Einstein recanted, publishing a new paper with his student Nathan Rosen in which he declared that gravitational waves could not exist. However, studies continued to prove Einstein's first version right: gravitational waves had to exist, and they had to be of two types, differing by an angle of 45°, depending on the orientation of the deformation axes. In practice, the two polarizations are the components that add together to form the gravitational wave.

The emission must come from the acceleration of a huge mass, so one possible source is a supernova, the explosion of a supermassive star. Its remnants give rise to a neutron star, although in some cases, the remnant is even denser and constitutes a black hole, itself a possible source of gravitational waves. The search for these waves is led by two international projects, the LIGO and VIRGO detectors, the first American and the second Italian–French: *"It is only with such big ears, maintained in the most absolute silence imaginable, that we may actually hear the music of the universe: the chirping of two black holes, the regular refrain of a pulsar, the high-pitched shriek of a supernova, the distant roar of the Big Bang"* [6].

On 14 September 2015 at 10:30:43 UTC, a black hole collision was observed by LIGO, and on 17 August, a collision was detected between two neutron stars 130 million light-years away in the Hydra constellation. The signal reached the two LIGO stations at Livingstone and Hanford and the VIRGO station with a delay of a few milliseconds, and NASA's Fermi satellite recorded the arrival of a gamma-ray burst, identifying the source in the galaxy NGC 4993.

8 The Universe Is Teeming with Aliens

The founder of the VIRGO project was an Italian scientist, Adalberto Giazotto. Like Galileo, he had a father who was a musician and musicologist. The announcement of the discovery of gravitational waves was accompanied by the Adagio in G minor by Tommaso Albinoni, but as Giazotto himself recounts, he would have preferred the second movement of Beethoven's String Quartet Op. 18 No. 1, because its scream of violence lends itself much better to the announcement of such a great discovery.

The physicist's father, Remo Giazotto, is remembered for his studies of Tommaso Albinoni, and in fact his monograph gave rise to the rediscovery of this composer. The famous adagio was completed by the musicologist himself, since Albinoni supplied only two themes and the bass line. But, when the conductor Ennio Gerelli performed the work publicly in 1949, it was attributed entirely to Tommaso Albinoni in the concert programs. The misunderstanding over the opera's authorship was resolved by the publisher Ricordi in 1956, when he recognized the work as the "Adagio in G minor by Albinoni and Giazotto." Remo introduced his son Adalberto to music when he was a university student, the eureka moment came when he heard Beethoven's 32 piano sonatas on a record by Glenn Gould: Giazotto writes: *"For me, Beethoven's depth and originality in music have an impact equivalent to general relativity and the other things I have been concerned with all my life in the field of physics. [...] The mystery of music, of how music becomes in our heads something that we appreciate and something that moves us, is basically the same mystery that allows us to understand how the universe is made"* [6].

This remark is very similar to what Einstein said about music and suggests a concept very close to that of the Pythagoreans. The search for gravitational waves began with Joseph Weber's idea of using resonant bar detectors, that is, bars that were sensitive to some well-chosen resonant frequency. Basically, these were large tuning forks that were activated in the presence of a "note" with a frequency between 700 and 1000 Hz, rather high-pitched sounds. It turned out that Weber's bars were not ideal for this type of research. A new technique was needed, and physicists opted for an L-shaped optical interferometry technique that would measure changes in the length of a laser beam whenever a gravitational wave passed through. In this way, a new era opened for physics and astronomy; the music of the spheres was no longer a philosophical speculation, but became a tangible reality.

In the words of Giazotto again: *The ancients thought the cosmos was pervaded by a harmonious and inaudible music, the music of the spheres. Our new instruments, listening to the movement of the planets and stars, will be able to give new meaning to this insight. But this as yet unheard melody will be an expression, not so much of the ideal shapes of spheres and polyhedra that were imagined in the past, but of the flexible and dynamic geometry given to us by Einstein"* [6].

The connection between astronomy, astrophysics, and music is also evident in the nomenclature of celestial objects. There is a yellow giant star of spectral type G6 III, called 18 Delphini, located about 238 light-years from Earth in the constellation Delphinus, and an exoplanet 18 Delphini b was discovered around this star in 2008. The planet was renamed Arion in memory of Arion of Methymna, an Ancient Greek citharist who is famous for having made structural modifications to the dithyramb, a song of the cult of Dionysus [7]. Herodotus relates that Arion wanted to make the journey to Sicily and Italy to promote his music. During the voyage, he managed to gain great wealth, and on his way back to the court of Periander, the Tyrant of Corinth, the sailors decided to rob and kill him. However, he was given the choice of suicide by throwing himself into the sea. Arion asked for a last wish: to sing one more time before he died [8]. The song he chose was in praise of Apollo. When he came to the end, he climbed onto the prow of the ship and threw himself into the sea.

On Apollo's recommendation, a dolphin which had been listening to his music took him on its back and carried him to the temple of Poseidon at Cape Tainaron. Unfortunately, in his haste to depart on his journey, he forgot to throw the dolphin back into the sea and it died stranded. When he returned to Periander, he told his story and the ruler gave orders to give the dolphin a proper burial. When the boat arrived, the sailors explained that Arion had died. However, he was hidden inside the dolphin's tomb and, when the king asked him to come out, the sailors were amazed. They were promptly sentenced to death at the dolphin's grave. According to the version of this story in *De Astronomica* by Hyginus, it is told that Apollo, impressed by Arion's prowess, carried the citharist and the dolphin into the sky, where they became two constellations, Lyra and Delphinus (see Fig. 8.7).

In 2014, the International Astronomical Union announced a competition to rename the star and its exoplanet in the constellation Delphinus. The host star was renamed Musica and its planet is now known as Arion.

The case of planet 18 Delphini b is not the only one. In 2004, astronomers at the Parkes Radio Telescope in Australia discovered a white dwarf, initially known as BPM 37093, in the constellation of Centaurus. As it was composed of carbon and crystalline oxygen, it was in effect a giant diamond. The star was subsequently renamed *Lucy* in honor of the Beatles' song *Lucy in the sky with diamonds*.

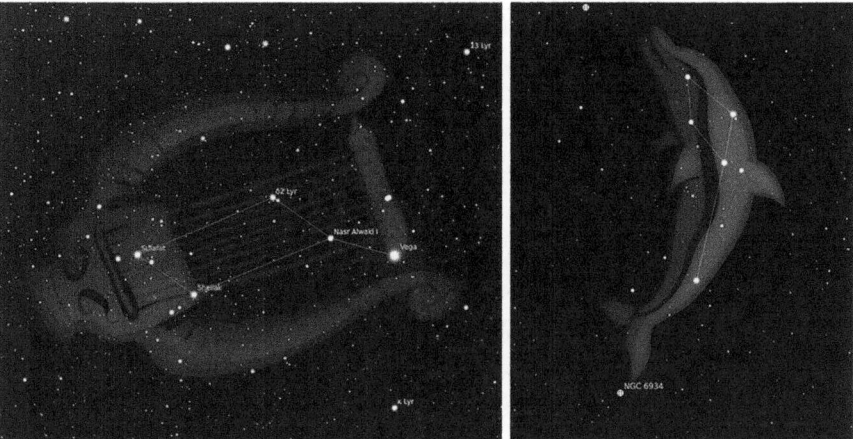

Fig. 8.7 The constellations Lyra and Delphinus. Public Domain

References

1. P. Ball, *The Music Instinct: How Music Works and Why We Can't Do Without It* (Brain Shot, Bodley Head, London, 2011)
2. C. Sagan (Jet Propulsion Laboratory, Retrieved September 23, 2010)
3. N.D. Tyson, *Astrophysics for People in a Hurry* (W.W. Norton & Company, New York, London, 2017)
4. Source NASA, www.nasa.gov
5. Italian Space Agency (ASI), *Missione Cassini, alla scoperta del sistema di Saturno* (Infokit, ASI)
6. A. Giazotto, *La musica nascosta nell'universo* (Einaudi, Turin, 2018)
7. G. Comotti, *La musica nella cultura greca e romana*, vol 1 (EDT, Turin, 1991)
8. E.M. Moomann, W. Uitterhoeve, *Van Achilleus tot Zeus* (SUN, Nijmegen, 1987)

9

The Songs of the Space Race

Interest in the music of the cosmos is also present in different forms in different arts, so the topic is not confined to its treatment in Classical Greece and science. In this regard, one can think of the miniature taken from the *Antiphonarium mediceum* of 1300, kept in the Biblioteca Medicea Laurenziana in Florence, which depicts Boethius' conception of Musica Mundana, Musica Humana, and Musica Instrumentalis (see Fig. 9.1).

In the twentieth century, interest in Musica Mundana spread beyond science to reach popular culture, built upon the historical substrate of events that shaped the twentieth century. The end of World War II and the beginning of the Cold War breathed a whole new life into the Pythagorean tradition. NASA was set up in response to the Soviet Union's first space activities, whence the space race became an extension of the Cold War on the scientific and technological front.

In 1968, Stanley Kubrik directed his film *2001: A Space Odyssey* in which music plays a dominant role, not only as a soundtrack, but also through the meaning the director himself attached to it. Music is used to establish a direct connection with the cosmos, but in contrast to the Pythagoreans who aimed for a deep level of research, filmmakers and musicians tend to try to capture the very meaning of the music of the universe. So, Kubrik's use of Johann Strauss' waltz "By the Beautiful Blue Danube" represents the cosmic dance of the planets in communion with the spinning of the space station. This is not a true music of the cosmos, but rather a symbolic representation mediated by our experience of the modern world. The same can be said of songwriters who have been inspired by human endeavors in space for some of their compositions.

Fig. 9.1 Musica Mundana, Musica Humana, and Musica Instrumentalis as illustrated in the *Antiphonarium mediceum*. Public domain

Words such as "stars," "planet," and "sky" are frequently used in the musical landscape, not to mention "Moon," the most celebrated celestial body in art of music. One only has to think of Beethoven's Sonata Op. 27 No. 2, the Moonlight Sonata, from Beethoven, Debussy's *Clair de lune*, the third movement of his *Suite Bergamasque*, and the Song to the Moon from Dvořák's opera *Rusalka*. And our very own natural satellite is still a source of inspiration for such famous songs as *Blue Moon, Moon River, Fly Me to the Moon*, and so many others.

The planets of the Solar System have found a place in musical discography with Gustav Holst's monumental work "The Planets," composed between 1914 and 1916 in the form of a seven-movement orchestral suite, in which each movement is dedicated to a planet: *Mars, the Bringer of War; Venus, the Bringer of Peace; Mercury, the Winged Messenger; Jupiter, the Bringer of Jollity;*

Saturn, the Bringer of Old Age; Uranus, the Magician; and Neptune, the Mystic. The only planet missing among them is our own.

NASA's space endeavors also influenced one of Italy's greatest composers, Bruno Maderna, whose work *Serenata per un satellite*, written in 1969 on the occasion of the launch of the ESTRO I satellite, was dedicated to the astronomer Umberto Montalenti, then director of the European Space Operations Center (ESOC) in Darmstat. The musical work is characterized not only by its structure, but above all by its geometric form; the score can be viewed as a modern portrait and belongs to the field of aleatoric music (see Fig. 9.2).

However, Maderna's work has a precedent in John Cage's *Atlas Eclipticalis* (see Fig. 9.3). The piece, composed between 1961 and 1962, took its cue from Antonín Bečvář's star catalog "Atlas of the Heavens," published in 1958. Cage superimposed his staves on the star charts and used the stellar magnitudes marked on the catalog to determine the duration of the notes. The work can be performed in many different ways, by up to 86 musicians.

On 20 July 1969, Neil Armstrong and Buzz Aldrin landed Apollo 11's lunar module on the Moon, while Michael Collins continued to orbit the Moon in the command module. The soundtrack for the BBC's live television coverage of the event was the song Moonhead, written by Pink Floyd specially for the Moon landing. For the journey, Armstrong took Antonín Dvořák's

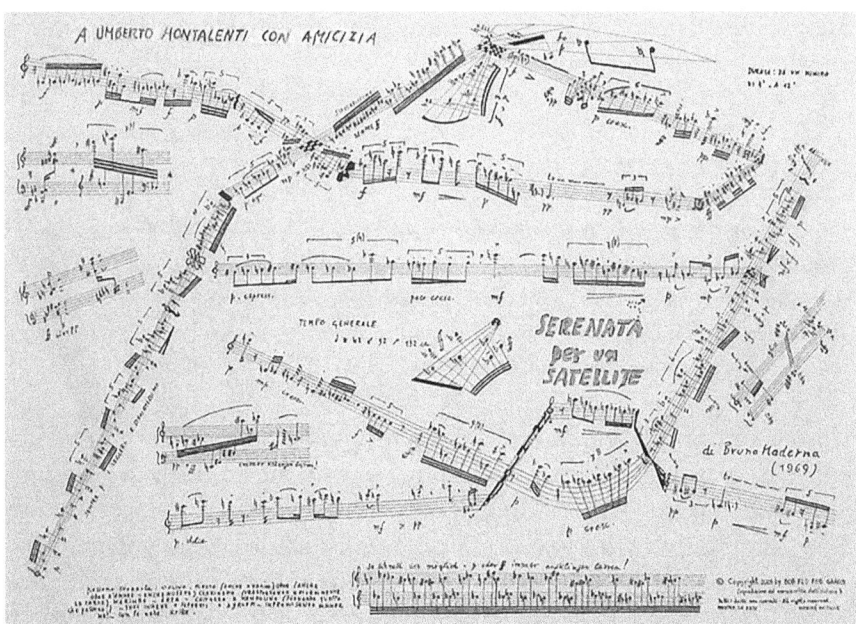

Fig. 9.2 Bruno Maderna's score, *Serenata per un satellite*. Photo by the author

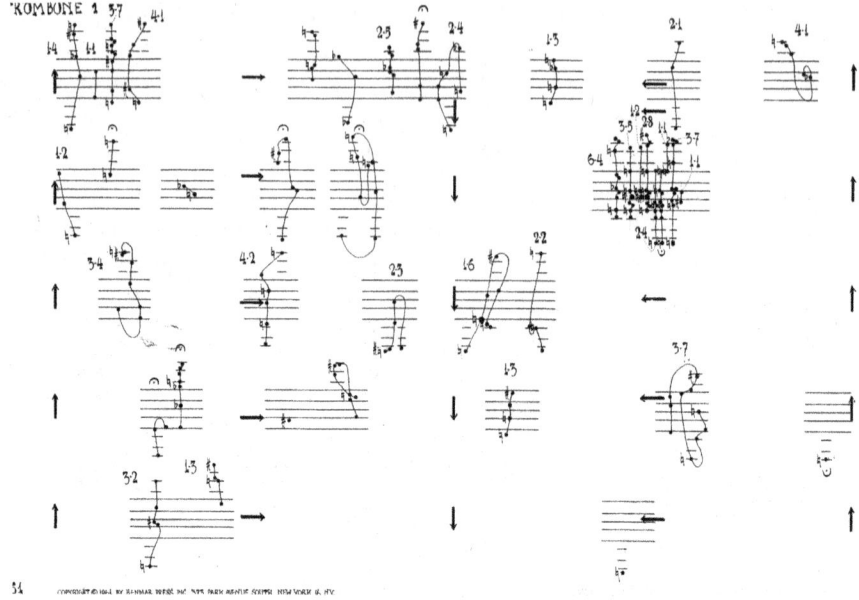

Fig. 9.3 John Cage's score *Atlas Eclipticalis*. Public domain

Symphony No. 9 "From the New World," so the Bohemian composer was effectively the first musician to land on the Moon. On a later mission, Apollo 12, astronaut Richard Gordon made the following remarks during a conversation with journalist Oriana Fallaci:

> [...] *There are many ways to die on the cross. It is a cross. But that damn trip works like this: you have to stay in the air and pick them up when they come back. [...] It worked with Neil, and it will work with Pete too. All I'm going to do during those thirty-six hours, though, is think about it. And when I have to sleep, I won't sleep. I would sleep better on the Moon, with them. Because I love them too much [...] I will listen to music, yes. Pete brought those cowboy ditties, but I brought Verdi, Puccini. It must be nice to listen to the chorus of Nabucco while you're all alone in the void and you can't even see the Earth. It must be a comfort, can you believe it* [1].

In the year of the first Moon landing, this momentous event inspired the Byrds, a band from California, to write a song with the explicit title *Armstrong, Aldrin and Collins*. Their musical genre became known as space country. Above all, 1969 was the year of David Bowie's album *Space Oddity*, which contained a song of the same name. The song was inspired by Stanley Kubrick's film *2001: A Space Odyssey*, released the year before. Bowie found the soundtrack interesting enough to include an extract from György Ligeti's

Atmosphere, a work that also featured in Kubrick's film. The lyrics of *Space Oddity* recount the launch and orbit insertion of a space module carrying Major Tom, a fictional character loosely inspired by real astronauts. The line "*Planet Earth is blue and there's nothing I can do*" brings to mind a remark made by Russia's first cosmonaut, Yuri Gagarin: "*From up here the Earth is beautiful, without borders or boundaries.*" However, Bowie's song does not seem to have a happy ending as something goes wrong on board and communication with Major Tom is cut off. This seems to be a reference to the early astronauts and cosmonauts who lost their lives in the first space launches. The song was revisited in 2013 by Canadian astronaut Chris Hadfield who, while in orbit on the International Space Station, made the first video clip in space (see Fig. 9.4). The recording of the voice and guitar, as well as the visuals, were made directly aboard the Space Station, while the arrangement was made here on Earth. The lyrics were revised by Colonel Hadfield, making them more appropriate to the present situation. He also gave them a more promising ending, since, after launching with the Soyuz spacecraft, Major Tom re-entered the atmosphere and landed safely back on Earth.

During the period when NASA was carrying out its Apollo program to take astronauts to the Moon, young bands, fascinated by these space missions, set in motion a musical revolution by creating the genre of *space rock*. Syd Barrett's Pink Floyd were among the early experimenters with *Astronomy Domine* (1967), a song in which Barrett extols the magnificence of the universe, alluding to the planets of the Solar System and their moons in the lyrics. The most significant song in the genre is *Interstellar Overdrive* (1967), an instrumental

Fig. 9.4 Chris Hadfield while filming the video clip of *Space Oddity* aboard the ISS. Figure credit: NASA/CSA

in which the otherworldly space or science fiction setting is provided by the use of distorted guitars and synthesizers. Even Jimi Hendrix, the famous guitarist, was fascinated by the Space Race, as reflected by the fact that he became a Star Trek fan. *Third Stone from the Sun* (1967) recounts the space voyage of an alien who wants to visit planet Earth; it is not a song, but rather a lyric recited to Hendrix's music. The visitor's journey ends with the destruction of the Earth. *Up from the Skies* (1968) is a sequel to the earlier song, as the alien visitor shows concern for the condition of the planet ruined by humankind. The Rolling Stones enter space rock with *2000 Light Years from Home* (1967), a song about a journey into deep space and precursor to Dik Dik's *Help Me*.

Another contributor to the theme of space rock was the British record producer and sound engineer Joe Meek who was fascinated by the various space programs and believed in the existence of life in our Solar System. In 1959, he began to elaborate his "Outer Space Music Fantasy," taking his cue from Holst's suite *The Planets* to create his first work, *I hear a New World* (1960), a concept album in which he depicted his idea of the cosmos in music. It is a twelve-track work in which he recounts the epic of three lunar peoples, the Dribcots, the Globbots, and the Sarooes. In 1962, he recorded *Telstar* with the band the Tornados. The song is clearly inspired by the deployment of the first telecommunications satellite, Telstar 1, owned by the American Telegraph & Telephone Company. To simulate the noise of a rocket launching, he blew into the microphone and reversed the sound of a toilet flushing.

Just as musicians look to space, there is nothing strange in the idea that astronauts may also be musicians. Indeed, it is not unusual to see musical instruments aboard the ISS. Flight engineer Edward T. Lu brought a keyboard (see Fig. 9.5), Commander Frank Culbertson had his trumpet with him (see Fig. 9.6), and French astronaut Thomas Pesquet was actually aboard the space station when he received a saxophone as a birthday present (see Fig. 9.7). In addition to these instruments, an accordion and an Australian didgeridoo also found their way on board, and astronaut Kjell Lindgren played *Amazing Grace* on the bagpipes (see Fig. 9.8).

Colonel Hadfield brought his acoustic guitar with him, but back in 1995 he had brought a specially made folding electric guitar aboard MIR, the Russian space station, so that he could perform a duo with ESA astronaut Thomas Reiter. The program included traditional Russian music and Beatles songs (see Fig. 9.9). Another of Hadfield's musical achievements was to have played in the first rock concert between Earth and space, when he performed live in the ISS Cupola with the musician Ed Robertson back down on Earth. However, there was a precedent. In 2011, to commemorate the 50th anniversary of Gagarin's feat, the astronaut Cady Coleman (see Fig. 9.10) played the

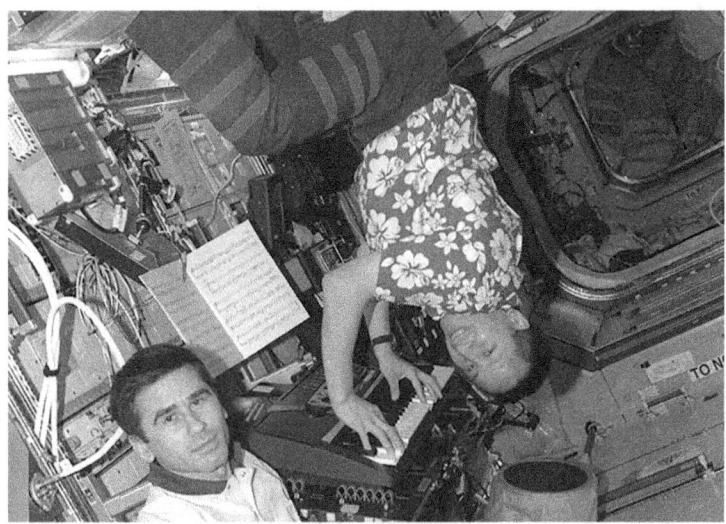

Fig. 9.5 Edward T. Lu's keyboard. Figure credit: NASA

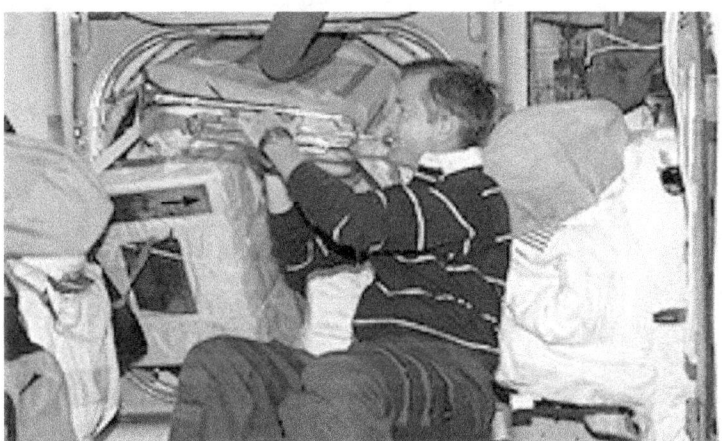

Fig. 9.6 Frank Culbertson playing the trumpet. Figure credit: NASA

flute live via satellite with Ian Anderson, the historic frontman and flautist of Jethro Tull. While the Golden Record was the first record launched into space, the first song in orbit dates back to 16 December 1965, when Gemini 6 astronauts Walter Schirra Jr. and Thomas P. Stafford (see Fig. 9.11), during re-entry to Earth, claimed they sighted an unidentified flying object on a north–south trajectory. According to them, the pilot was wearing a red suit. They thus began playing Jingle Bells with a mouth organ and bells. In 2018, forty years after its release, the song *Spacelab* by Kraftwerk was performed by

Fig. 9.7 Thomas Pesquet with his sax. Figure credit: ESA/NASA

Fig. 9.8 Kjell Lindgren playing bagpipes. Figure credit: NASA

the astronaut Alexander Gerst, who was also from Germany, like the historic group. On this occasion Gerst played a virtual synthesizer with the pioneering musicians of electronic music in a concert commemorating the 1978 album *The Man Machine* in Stuttgart (see Fig. 9.12), introducing the five notes of *Close Encounters of the Third Kind*.

Italy can also boast a musical record. In 2019, the astronaut Luca Parmitano (see Fig. 9.13) dedicated *Feliz Navidad* by José Feliciano to the whole world. He played guitar and was accompanied by astronauts Jessica Meir (piccolo), Christina Koch (maraca), and Andrew Morgan (whistle). Jessica Meir was not

Fig. 9.9 Chris Hadfield on electric guitar and Thomas Reiter on folk guitar. Figure credit: NASA

Fig. 9.10 Cady Coleman on the flute. Figure credit: NASA

new to such performances; she had already played the main theme of Star Wars in the International Space Station.

Returning to David Bowie, the artist did not only give the world *Space Oddity*; he also produced *Life on Mars*, *Starman*, and *Hallo Spaceboy*, the songs that accompanied us through the years of space launches and the Apollo missions to the Moon. And like Bowie, Elton John also became fascinated by the dream of man in space. According to writer and scientist Isaac Asimov: "*If ever the ancient belief of the "music of the spheres" could be said to have come literally*

Fig. 9.11 Walter Schirra Jr. and Thomas P. Stafford. Figure credit: NASA

Fig. 9.12 Alexander Gerst and the Kraftwerk. Figure credit: NASA

true, it was in that hum of hyperatomics that was the very essential of space flight" [2]. Elton John produced *Rocket Man* in 1972. It was actually inspired by one of Ray Bradbury's short stories included in the collection *The Illustrated Man*. Bradbury is best remembered as the author of *Fahrenheit 451* and *The Martian Chronicles*. The song was published only three years after the first Moon

Fig. 9.13 Luca Parmitano playing the guitar and kazoo on the ISS. Figure credit: NASA

landing, but Elton John was already imagining what it might be like to live on Mars: *"[…] Mars ain't the kind of place to raise your kids; in fact it's cold as hell and there's no one there to raise them if you did […]* [3].

The space race was not only studded with success. Indeed, its failures proved to be real tragedies. Many astronauts lost their lives in these attempts to reach the stars, and some of their names remained classified in the archives of the Russian and American space agencies. The brothers Achille and Giovanni Judica-Cordiglia from Turin claimed to have intercepted communications between Russian cosmonauts and mission control. These voices from the darkness were shouting and calling for help. In the transcripts of the recordings, these words are heard:

"I'm hot! I'm hot! Done. It's so hot! I'm hot! I'm hot! There are flames on board. What? I can see the fire. There's a fire. There's a fire. Hot. Hot. Thirty-two! Thirty-two! Forty-one! Forty-one! Am I crashing? Yes, yes. Hot. Am I re-entering? Am I re-entering? Yes, I'm listening. Hot. It's all over!" This is attributed to a Russian cosmonaut during a flight that ended in tragedy. However, it must be remembered that the days when the Judica-Cordiglia brothers were listening to signals from space was also the era of the Cold War, and since there are no official records regarding the astronauts alleged to have disappeared in space, a piece of disinformation by Russian, American, Italian, or whatever other intelligence services cannot be ruled out. But it is not important whether these recordings are true or false; what matters is that they fueled a certain curiosity and literature, and testify also that music did not remain insensitive.

While enthusiasm for *"one small step for man, one giant leap for mankind"* inspired many songwriters, space tragedy also became a source of inspiration. In 1974, the Italian band Dik Dik released the song *Help me*, whose lyrics tell the story of the astronaut McKenzie who loses all communication during his journey to Jupiter. Apparently, his spaceship was hit by a meteorite and *"of McKenzie nothing remains, just a tape that recorded the voice of a frightened man"* [4]. However, critics say the lyrics were inspired by David Bowie's *Space Oddity* and that the idea of the trip to Jupiter was borrowed from the voyage of the *Discovery* in Stanley Kubrik's film *2001: A Space Odyssey*.

In an interview with the author [5], Pietruccio Moltalbetti, founder of Dik Dik and author of the song, says he did not know the song by David Bowie and therefore had no inspiration from Space Oddity. Rather, he says: *"The song was a choice, and it came about because I was intrigued by the search for something we may never know. Jupiter is our sentinel, the universe is boundless, and I believe there are alien life forms elsewhere […] I have traveled the world and seen disturbing things. I have been with the Dogon people, who claim to come from the star Sirius B and worship it […]."* Although the voyage in Kubrik's film is also to Jupiter, this may just be due to a simple analogy since *"Jupiter is difficult to get to compared to Mars or the Moon. It is a matter of reaching the last frontier, and now that frontier is space itself […]."* There was one last question: whether a disaster narrative was deliberately chosen and whether McKenzie was the name of a real astronaut who died on some mission: *"McKenzie is a classic, purely invented name. Before we got to the Moon, there were several space disasters, just as many people lost their lives trying to climb Everest or discover America. The song draws from history and historical disasters […]."* Moltalbetti concluded: *"We will get to Mars, but there will be major disasters. Before achieving such a goal, there is always some sacrifice."*

References

1. O. Fallaci, *La luna di Oriana* (Rizzoli, Milan, 2018)
2. A. Adamo, *Pianeti tra le note* (Springer-Verlag Italia, Milan, 2010)
3. E. John, R. Man, *From the Album "Honky Château"* (MCA Records, DJM Records, 1972)
4. From the song *Help me* by Dik Dik
5. From an interview with Pietruccio Moltalbetti on 1 November 2018

10

Space, the Music of the Spheres, and Movies

In space no one can hear you scream [1]

Talking about the music of the spheres through film provides a way to explain certain concepts by analogy. It seems obvious, even superfluous, to point out that directors and screenwriters simply apply different kinds of music according to the criteria of the narrative, so there is no real correspondence between the Musica Mundana born from philosophy and the Musica Mundana in the world of cinema, since the latter is only an artistic evocation of the former.

The soundtrack of a film is not limited to a simple musical soundtrack, but includes the entire soundtrack of the scenes, from background noises and diegetic music which are directly involved in the action, to extra-diegetic music, all of which interact with the scene to become part of the soundtrack. A system like this can evoke sensations of huge open spaces, broad expanses of land or ocean, and skimming flights over water or through interstellar expanses, creating sound emotions encoded in the collective imagination.

Stanley Kubrick's interplanetary space is drawn from the works *Lux Aeterna, Atmospheres, Adventures,* and *Kyrie* by György Ligeti, the dance of the spaceships and of the planet Earth is characterized by the Johann Strauss Jr. waltz *By the Beautiful Blue Danube*, and "the dawn of man" is remembered by the introduction of the symphonic poem *Also sprach Zarathustra* by Richard Strauss. Although in Kubrick's case, we cannot speak of a Musica Mundana in the strict sense, because there is no real intention to create a connection with the Pythagorean or Boethian theory, when the notes of Richard Strauss echo around a concert hall, our minds fly straight to Kubrick's deep space.

In all likelihood, Tim Burton's Martians in the movie *Mars Attacks* (1996) had not intercepted the Golden Record of the Voyager probes, so when they come to Earth to exterminate humankind, they become victims of music. The song that ends their attack is *Indian Love Call*, a country song sung by Slim Whitman in 1952, while the finale is by Tom Jones, a survivor of the invasion, who sings *It's Not Unusual*. According to Oliver Sacks, we are in the presence of a "seductive" music which, as he puts it, can be so intoxicating to some and so annoying to others [2].

Once again, the science–music dualism has to be taken in an artistic way. Even though one might advance the hypothesis that the end of the Martians was due more to the frequency of Whitman's notes than to the song itself, in a more sarcastic vein, the opposite might actually be true. The interesting thing here is the idea of using music as a lethal weapon against the invaders. This is reminiscent of the song *Lili* Marleen, which was banned by the German propaganda minister Goebbels because it was too "defeatist," and instead of inciting soldiers to fight, encouraged them rather to lay down their arms and return to their families. In both cases, one could say that the army of the Martians and the army of the Germans would have been defeated by music.

The film that comes closest to the relationship between music and the cosmos is *Contact* (1997), directed by Robert Zemeckis and based on a novel by Carl Sagan, in which a young astronomer, played by Jodie Foster, while listening to the "music of the cosmos," intercepts an alien signal. Indeed, the work she is engaged in for Project Argus uses radiotelescopes to listen to the signals that come from deep space, a practice that already approaches the idea of Musica Mundana in the Pythagorean sense. The movie begins "[…] *with a suggestive race through space of something invisible that brings with it the sounds, the voices, the noises of our radio emissions, from the first, timid notes sent into the ether, of Hitler's speeches heard by a terrified and incredulous world to the latest powerful emissions from our radio and television antennas*" [3].

The signal that gets picked up by the main protagonist is clearly reminiscent of the Wow! signal, with talk of harmonics and prime numbers in the sequence in which it is analyzed. However, this signal was already famously alluded to even before Zemeckis' film, in the opening episode of the second season of the television series *The X-Files*, which aired in the USA in 1994. In this episode, FBI agent Fox Mulder (alias David Duchovny) stumbles upon the "alien" sequence. After learning about the Wow! signal, he goes to the Arecibo radiotelescope, where he meets the Puerto Rican Jorge, who, terrified, draws him a picture of the extraterrestrial with whom he claims to have had a close encounter.

In *Contact*, this signal is presented as a sound that carries with it a message in the form of prime numbers and a television signal. Later, it gets translated into the technical details for building a machine that can take them through space-time to meet the civilization that sent it. The audio signal is heard by Ted, a blind astronomer, who cannot see the signal as a graph, in the form of waves, but can only hear it. This expedient is important because, once again, it connects the scientific aspect with the Pythagorean idea of music, and in particular, its numerical form. In this sequence the sound, or rather the transmission frequency, encodes a series of prime numbers carrying a message, and in the Pythagorean tetraktys, the prime numbers are 1, 2, and 3. In another sequence, there is a meeting with the presidential staff to discuss the meaning of the message discovered by Ellie Arroway (alias Jodi Foster):

– Arroway: *All I'm saying is, this message was written in the language of science – mathematics – and was clearly intended to be received by scientists. If it had been religious in nature it should have taken the form of a burning bush, or a booming voice from the sky...*
– Palmer: *But a voice from the sky is just what you say you've found.*

In fact, the idea of a "message from the sky" reminds us of the title of Galileo's work on astronomy, *Sidereus Nuncius*, which translated literally means "message from the stars." The film is based on the novel of the same name published in 1985 by Carl Sagan, astronomer and founder of the Search for Extraterrestrial Intelligence (SETI) project. When he wrote the novel, Sagan wanted the scientific aspects to be as plausible as possible, not fake scientific. For this reason, the astronomer turned to his friend Kip Thorn [4], a leading expert on general relativity, for advice on how to make the story more credible.

The idea of communicating with alien worlds was also put forward by Nikola Tesla, who said: "[…] *with the novel means, proposed by myself, I can readily demonstrate that, with an expenditure not exceeding two thousand horsepower, signals can be transmitted to a planet such as Mars with as much exactness and certitude as we now send messages by wire from New York to Philadelphia. […] there would be no insurmountable obstacle in constructing a machine capable of conveying a message to Mars, nor would there be any great difficulty in recording signals transmitted to us by the inhabitants of that planet, if they be skilled electricians*" [5]. So, the idea of using sound as an extraterrestrial signal involves an obvious Pythagorean approach in its musical conception.

The aliens in *The War of the Worlds* (2005) directed by Steven Spielberg destroyed humans with large tripods that announced themselves to the sound of a minor third (e.g., C-E flat). This is rather the opposite to what happens in *Mars Attacks*, where the Martians were destroyed by music, because here they use sound to make themselves seem more threatening. Once again, we cannot speak of Musica Mundana, but rather of Musica Instrumentalis. However, the idea of attributing a motif to an alien machine, albeit comprising only two notes, shows the director's need to reconnect the idea of Giordano Bruno's infinite number of worlds with the existence of a Musica Mundana, even if only in the artificial sense of the term. This remake of the 1953 film of the same name and of the 1938 radio drama is very different from the musical idea that Spielberg and the composer John Williams expressed in 1977 with *Close Encounters of the Third Kind*, but the use of the musical theme to underline what is alien to us, as beings from another world, remains evident and fundamental.

In the *Star Wars* saga, the music goes beyond the soundtrack, becoming a player in the film and suggesting the existence of a Musica Instrumentalis in other civilizations. This is the case of Figrin Da'n and the Modal Nodes, a band of the Bith race from the planet Clak'dor VII that performs in spaceport cantinas rather like pubs. They appeared for the first time in *Star Wars*, the first episode of the saga released in 1977, later renamed *Episode IV – A New Hope*. The theme was composed by John Williams and sounds rather "earthly," like the blues sound of the band in Jabba's Palace in *Return of the Jedi* (1983), but in *Episode III, The Force Awakens*, the work that young Anakin and Senator Palpatine attend sounds very "alien." There are many other examples in *Star Wars*. Aliens performing music are a constant throughout the saga. It suffices to mention the Great Municipal Band in *Episode I – The Phantom Menace* (1999), where a band of Gungans from the planet Naboo perform in a parade to celebrate victory in the war against the Galactic Empire.

In stark contrast, in Luc Besson's film *The Fifth Element* (1997), we can appreciate an alien opera singer performing the aria "Il dolce suono" from Gaetano Donizetti's opera Lucia di Lammermoor. In the original text, we find a line that reads: "*Un'armonia celeste, dì, non ascolti?*" (A heavenly harmony, tell me, do you not hear it?) The heavenly harmony relates here to the love song, but the "alien" scene in the film brings us back to an example of Harmonia Mundana that is even more meaningful because it is the same harmony sought by Pythagoras and the Pythagoreans, and now, symbolically, sung by an alien.

In the world of television series, it is worth mentioning the saga *Star Trek: The Next Generation*, because it contains so many musical references. Just as real astronauts brought their musical instruments aboard the International Space Station, so too Commander Riker carries his trombone around the cosmos, while Data, the android, performs with his "mother" in a violin duet, and we see him playing again in a string quartet. Captain Picard performs on an "alien" flute in the episode "The Inner Light;" here, he relives the events of a vanished civilization, of which only the flute remains as a concrete testimony. In the original series *Star Trek*, Spock is a lute player and Uhura is a singer, while in *Star Trek: Voyager*, the holographic doctor performs opera and Lieutenant Kim is a clarinet player. In the episode "Penumbra" of *Star Trek: Deep Space Nine*, Lieutenant Worf sings Klingon opera, and finally, in the same series, we discover that even Commander Sisko seems to know how to play the piano.

The space of science fiction is full of musicians, performances, and cosmic music in both the Pythagorean and instrumental senses, and to paraphrase Ed Robertson: "*Turns out in space lots of people can hear you scream*" [6].

References

1. Tag-line from the movie *Alien* by Ridley Scott, 1979
2. O. Sacks, *Musicofilia: Tales of Music and the Brain* (Picador, London, 2018)
3. A. Adamo, *Pianeti tra le note* (Springer-Verlag Italia, Milan, 2010)
4. K. Thorn, *Black Holes and Time Warps: Einstein's Outrageous Legacy* (W.W. Norton, New York, 1994)
5. N. Tesla, *Talking with the Planets* (Collier's Weekly, 9 February 1901)
6. *Turns Out in Space Lots of People Can Hear You Scream*, title for the album made by astronaut Chris Hadfield, suggested in jest by the musician Ed Robertson

11

Close Encounters of the Third Kind

Where did these sounds come from? [1].

It was on 16 November 1977 that American cinemas gave the first showing of *Close Encounters of the Third Kind*, directed by Steven Spielberg, just five months after the release of George Lucas's film *Star Wars*. This was the same year as the launch of the two Voyager probes and the discovery of the Wow! signal. All things considered, 1977 was therefore a bumper year for the world of astronomy and science fiction films. In the years up to 1986, we can speak of a genuine UFO mania, with a proliferation of space-based science fiction films and a sometimes desperate scientific and pseudo-scientific search for a much desired alien contact. Some historians argue that ufology can be seen as a sort of alter-ego of communism and the Cold War, wherein the concept of "aliens" would have little in common with beings actually coming from other worlds. It would rather be an embodiment of the Western fear of communists, who would be viewed as "aliens" in the democratic West. The 1953 film *The War of the Worlds*, based on the novel by H.G. Wells, and the later radio adaptation of the same by O. Welles, tells of the arrival of Martians who seek to conquer the Earth and destroy humankind, but a clear criticism of European colonialism underlies the author's narrative. Whatever the explanation for this trend, it remains a fact that it was the most prolific decade for the space movies. Along with successes such as *Star Wars* and *Close Encounters*, other films that came out over this period were *Alien* (1979, Scott), *Star Wars: The Empire Strikes Back* (1980), *E.T. the Extra-Terrestrial* (1983, Spielberg), *Star Wars: Return of the Jedi* (1983), *Aliens* (1986, Cameron).

The space movie craze was inaugurated by Stanley Kubrick's *2001: A Space Odyssey*, which was released in 1968, a year before the Apollo 11 Moon landing. This coincidence and the fact that Kubrick was able to work with NASA for the production of his film have fueled theories that the Moon landing was actually a fake and that the images aired in the live worldwide broadcast in 1969 were shot by the director on a film set that was specially designed for the occasion. After the "small step for man," the lunar missions lost their appeal, but at the same time the interest in ufology and the search for extraterrestrial life was growing fast, and the films by Spielberg and Lucas arrived just in time to keep alive the public imagination as regards space travel and contact with new civilizations.

The two directors were not at all sure how successful they would be with their respective films. Lucas was fascinated by Spielberg's project and having just pulled off the cult hit *American Graffiti*, he didn't think he could achieve the same result. So, when he went to the set of *Close Encounters*, because he was sure that *Close Encounters* would do much better than *Star Wars*, he decided to make a bet with his colleague Steven, namely, to exchange 2.5% of the box office takings of each of the two films. But things turned out quite differently. *Star Wars* grossed 460 million dollars in the United States alone, beating Spielberg's film by a wide margin, and allowing its director to collect a tidy sum from the bet with his friend Lucas.

The idea for *Close Encounters* goes back to an early Spielberg production, *Firelight*, a movie that cost only $500 and was shown only once, in a cinema in Phoenix, Arizona, in 1964. The film tells the story of two scientists investigating some strange lights in the sky in Arizona. It combines the family problems of one with the research activities of the other, who is intent on convincing the CIA that extraterrestrial life does not exist. In the end, three aliens turn up and reveal their real purpose: to transport the imaginary city of Freeport to their planet to create a human zoo. This film already contains all the elements of the 1977 film, including the relationship between Roy and his wife and the arrival of the mother ship from which the little aliens emerge.

The plot of *Close Encounters* is very simple. Roy is an employee of the electricity company. He has a house, a wife, and three children. His life is jogging along in quite normally until, due to a break in the power supply, he is called out to solve the problem. On his way in the company van, he stops at a level crossing to consult a road map, and in the rear-view mirror, he spots two lights approaching from behind. He motions them to go ahead, but when the car behind him goes past, two more lights appear in the rear-view mirror. Again, he signals them to pass, but the lights suddenly brighten and his car is illuminated by a light so intense that, as soon as he leans out of the window to see what it is, half his face gets burnt. The light generates not only heat, but also an intense magnetic field that turns on all the electrical systems, including those of the level crossing. Roy has had his first encounter with a UFO.

From that moment on, his life changes. He starts going crazy. In everything he does, he has the feeling that he must give some kind of explanation for that event, so he starts to sculpt the shape of a mountain, first with mashed potatoes and then with soil and other materials he finds in his garden. Roy seems to have gone mad. The neighbors grow quite afraid of him and his wife runs away, taking their children with her. When Roy eventually comes back to his senses, he stumbles across a TV show that brings enlightenment. In the background he sees the very mountain that has been in his mind's eye: the Devils Tower in Wyoming. All his doubts vanish immediately.

Before the events of Roy's story, we meet the scientist Claude Lacombe (played by François Truffaut) and his interpreter David Laughlin. Lacombe was at the Montsoreau conference, where he spoke about the case of a UFO sighting in 1966. He and Laughlin are investigating the mysterious reappearance of Squadron 19, which disappeared in the Bermuda Triangle in 1945 and then turned up again in the Sonoran Desert, with the planes in perfect working order, but no pilots.

At the same time, in Muncie, Indiana, three-year-old Barry Guiler is woken up by strange lights that make his toys come to life. However, he is not frightened, but rather intrigued, and runs outside to chase after the lights, while his mother Jillian also notices the lights and runs after her child to stop him. Later, the mysterious lights reappear in Barry's house, but this time the child is kidnapped. Jillian develops the same obsession with the alleged UFO message that ruined Roy's life. The two eventually meet at the Devil's Mountain military base camp where a landing strip has been set up to facilitate communication with the aliens. Roy joins the group of volunteers who will board the spaceship, while all the abductees return, including the pilots of Squadron 19, and Barry can finally hug his mother once more.

In the film's plot, the most significant role is that of Lacombe, who discovers the aliens' method of communication thanks to the "Indian chant," consisting of a simple 5-note melody. Lacombe translates this melody into a visual form of communication, using the method developed by Zoltán Kodály to teach music to children, and uses it to present his research at the conference.

Music plays a key role in this film. In fact, many aspects of the film can be treated through the relationship between the staging of the film and the music. There are two fundamental aspects, the first related to the method of communication and the second to the universal nature of Musica Mundana.

Fig. 11.1 The sequence of five notes in the key of F major

Fig. 11.2 The sequence of five notes in the key of C major

The recurring melody in the film, also performed by Barry on the xylophone, consists exclusively of five notes. John Williams, author of the soundtrack, stuck scrupulously to the instructions given to him by Spielberg, who wanted a five-note melody that would be the equivalent of the word HELLO. On the face of it, the choice may seem simple enough, but in reality, Williams tried thousands of combinations before finally making what appears to be a completely arbitrary choice, but one that proved to be profound in meaning.

The five notes were G - A - F - F^{8va} C (see Fig. 11.1).

But in the "learning sequence," the final scene of the film, they are played in the key of C major (see Fig. 11.2).

These five notes are very special because they are closely related to the Pythagorean world and Musica Mundana, both for their numerical features and for what they represent.

For convenience, it is more useful to analyze the sequence in the key of C major, because the interval and numerical ratios are the same in any other key, but they would be less familiar to the reader. From a numerical point of view, we can say that the sequence is built on the II, III, I, VIII, and V degree of the C major scale (see Fig. 11.3). We can therefore relate the sequence of the five notes to a numerical series:

$$2,3,1,8,5$$

11 Close Encounters of the Third Kind

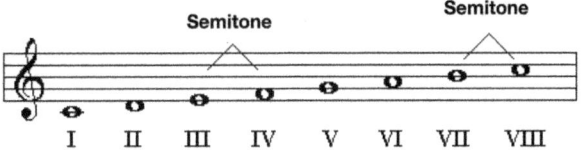

Fig. 11.3 C major scale showing the eight degrees

Considering that the octave has the same sound but double the frequency of the first degree of the scale, we can rewrite the numerical series in the following way:

$$1, 1, 2, 3, 5, 8$$

This series should look familiar. In fact, these are the first six numbers of the Fibonacci series, in which each number is the sum of the previous two. It continues thus:

$$1\ 2\ 3\ 5\ 8\ 13\ 21\ 34\ 55\ 89\ 144 \supset$$

Another characteristic is that, excluding the octave due to the fact that it can be traced back to the fundamental note of the scale, they are all prime numbers.

If we divide the number 8 by the number 5, we get a rather peculiar result:

$$\frac{8}{5} = 1.6 \quad (11.1)$$

This is an approximation to a fascinating result. If we consider the twelfth number of the Fibonacci series and divide it by the previous one, we get a much more precise approximation to that result:

$$\frac{144}{89} \oplus 1.618 \quad (11.2)$$

As we proceed along the series, the ratio of each number to its precedent tends to the value of the golden section, defined exactly by

$$\varphi = \frac{1+\sqrt{5}}{2} \approx 1.618 \quad (11.3)$$

So, Williams' five notes represent the golden ratio.

Looking at the ratio of the third term to the second, viz.,

$$\frac{3}{2} \qquad (11.4)$$

we can see a further analogy, this time with Kepler's music of the spheres. Indeed, in his third law of planetary motion, the astronomer understands that the ratio between the cube of the semi-major axis of a planet's orbit and the square of the period of revolution is constant:

$$\frac{a^3}{T^2} = \text{Const} \qquad (11.5)$$

If we remove the alphabetical symbols from the expression, we get

$$\frac{3}{2} \qquad (11.6)$$

This is the exact value of the fifth note found by Pythagoras with his experiments on the strings (see Fig. 3.3).

The five notes chosen by John Williams are thus full of metaphysical meanings, shedding their intrinsic musical value and acquiring what Iannis Xenakis [2] would have called a meta-musical value. The aliens communicate with earthlings through these five notes. The language becomes universal because it is no longer based on the perception of music, but on the mathematical meaning that it brings with it. Communicating with these five notes really means communicating with the universe, and these aliens communicate using music.

There is a historical precedent for the use of five notes. It occurs in the final trio *Hab' mir's geblot* in Act III of Richard Strauss' opera *Der Rosenkavalier*, first performed in 1911. The five-note melody is played by the first trumpet (see Fig. 11.4).

In this case, however, there are some differences. In fact, the numerical ratios are 2, 3, 7, 5, and 1, all prime numbers, and if we consider G flat as the

Fig. 11.4 Excerpt from the orchestral score of Act III of the opera *Der Rosenkavalier*

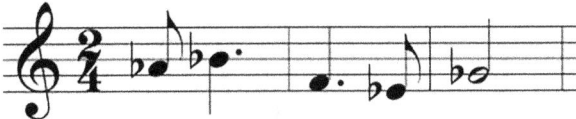

Fig. 11.5 The five notes from Strauss's opera

fifth, then we will be in the scale of C flat major (see Fig. 11.5); but in reality, Strauss was not thinking of using the five notes in the sense sought by Spielberg and Williams, although it could be considered as one of the possible combinations among those proposed for the film.

The idea of including the music of the spheres in a stage work dates back to 1589, when Ferdinando de' Medici commissioned a play by Girolamo Bargagli for the occasion of his marriage with Christine of Lorraine [3]. The work was called *La Pellegrina* and contained six musical interludes played by members of the *Camerata de' Bardi*, namely Giulio Caccini, Giovanni de' Bardi, Jacopo Peri, Emilio dei Cavalieri, Antonio Archilei, Cristofano Malvezzi, and Luca Marenzio.

The first interlude is called *Armonia delle sfere* (Harmony of the spheres). It consists of six pieces, the first of which, entitled *Dalle più alte sfere* (From the highest spheres), comprised a text by Giovanni de' Bardi, set to music composed by Antonio Archilei. This interlude was dedicated to the neo-Pythagorean concept of Harmonia Mundi and opened with a song sung by Armonia Doria, played by Vittoria Archilei, sitting on a cloud with Rome in the background and accompanied by two hidden guitars and a lute, all string instruments, as if to recall the experiments carried out by Pythagoras. In the second part of this interlude, the Fates, Astrea, and the seven planets appear, in a dialogue with ten sirens, set to music by Malvezzi [4].

Williams' contributions to several films are clearly analogous to the interludes in *La Pellegrina*. He used this five-note system in two other works for which he wrote the music, namely, Star Wars and E.T. the Extra-Terrestrial. The first film was released in the same year as Close Encounters, so it seems quite plausible that Williams used the same modus operandi. In fact, the main theme of Star Wars is built on the same interval relationships between the notes (see Fig. 11.6). And once again, we find the Fibonacci sequence, because we have the fundamental sound C, its octave, the second (D), the third (E), and the fifth (G), hence, 1, 1, 2, 3, 5, 8 but this time in a different arrangement.

The same happens in the main theme of E.T. (see Fig.11.7).

Fig. 11.6 The main theme from Star Wars

Fig. 11.7 The main theme from E.T

A composer's creative ability is measured not only by their melodic inventiveness, but even more so by their skill in producing variations, something that was so important to Beethoven and Mozart. Williams is a neo-Pythagorean, and he demonstrates this in all three of the space-inspired films discussed here.

The origin of the five notes is revealed in the scene in which the Indians, while singing the song, are asked: "*Where did those sounds come from?*" The answer is made clear to all in a single frame of the film: everyone raises their hand and points to the sky. The Indians intone "Aaya Re! Aaya!" which in Hindi means "he has come."

During the filming in India, Spielberg *"hired a local Indian choral leader to stand on a hill and lead the three thousand extras in the famous five notes with which we will eventually communicate with the Mothership"* [5], but with each take, the choir leader mispronounced the five notes. The production assistants sang the melody in his ear, but he kept getting it wrong, until he *"removes a pair of thick eyeglasses and a pack of Camels from [a religious prayer gourd]. He lights up and nervously studies his music. [...] He never does get the notes right, and all three thousand extras spend several hours following him in the incorrect response. We're falling behind schedule. Many of the extras are religious Moslems, and during their lunch break they have gone to pray at a nearby mosque. [...] We spend another hour trying to get a horde of natives to point straight up to the sky and yell. Take after take, one native points his finger in the wrong direction, or another takes his hand down before Steven is finished filming"* [5].

This story, told by Bob Balaban, the actor who played Lacombe's translator David Laughlin, is quite evocative and is accompanied in Balaban's book [5] by a photo from "behind the scenes" which shows the Indian's pointing toward the sky. Spielberg, upon returning to Hollywood, still had doubts about the

Fig. 11.8 The notes actually sung by the Indians

Fig. 11.9 The notes of the Indian chant rearranged on the score according to the five-note series

five notes sung by the Indians. In fact, the Indian chant sounds very different from the five notes (see Fig. 11.8).

The reason is very simple. The Indian musical system is very different from the tempered Western musical system, so while it is easy for a Westerner to intone the five notes, it certainly is not for an oriental. However, the choir leader's efforts were not in vain. Although the melody does not follow the same sequence of intervals required by Williams and the key is different, the mathematical relationships between them remain the same. So, we still find the series 1, 2, 3, 5, 8, and rearranging it rather like a musical anagram, we retrieve the five notes (see Fig. 11.9).

Spielberg's concern about the end result seems justified if it had been a European or American song, but it would have been difficult to get the performance wrong. In the Indian case, one can justify such an error as a musical "translation" of the five notes into a non-Western system, and this would make sense, since it is a communication between aliens and different world populations. Indeed, not all peoples speak English, and in the same way, not all peoples have the same musical system. What does not change is the mathematical relationship between the sounds, and this is enough to justify such an error in the film from an aesthetic point of view.

Lacombe behaves as a perfect anthropologist, or rather ethnomusicologist. In fact, he records the Indian song on magnetic tape and reveals its secrets at a conference. He realizes that the song consists of five notes and translates the sounds into a language of non-verbal communication through the use of hand signs. The idea is clearly expressed in the film by a conversion chart from notes to signs upon which is written "Zoltan Kodaly a visual learning aid." The

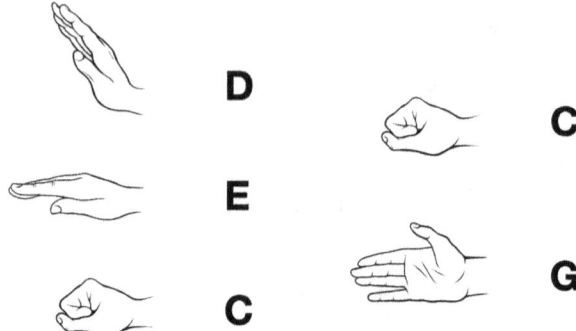

Fig. 11.10 Sequence of hand signs from the Kodály method as used by Lacombe

method used by Lacombe is Kodály's method for teaching musical notes to children (see Fig. 11.10).

Kodály (1882–1967) was a Hungarian composer, but he dedicated his life to the search for ancient musical traditions in his country, becoming a pioneer of ethnomusicology. He was also interested in musical education, writing music for didactic purposes, together with many theoretical works. Among the latter, the most famous is the Kodály method, based on the principle of solmisation, a system invented a thousand years ago by Guido D'Arezzo to instruct singers through the use of chironomy.

During a trip to England, the Hungarian composer was able to observe a new musical practice in use in schools to teach choir singing. The method of notation was devised in the early nineteenth century by Sarah A. Glover and perfected by John Curwen, and it was in fact based on the practice of chironomy, so he had the idea of adapting the hand gestures, while acknowledging Curwen in the preface to his account of the method.

Lacombe thus decided to communicate with the aliens through a gestural language combined with sounds which are for their part carriers of mathematical information. Moreover, we even see him send a musical message into space using a Yamaha SY-2 keyboard, playing the five tones and receiving in return the coordinates of the landing site.

At the end of the conference, all the scientists stand up to applaud, and at the back of the room, we see the first people who will later board the alien spacecraft. In the final scenes, we note that there are eleven of these, who are joined by Roy, the only one to be chosen to board the ship by the aliens themselves. So, the group consists of a symbolic number of people, corresponding to the twelve degrees of the tempered musical scale, the number twelve as in Schoenberg's twelve-tone technique, and the square root of the number 144, the twelfth number in the Fibonacci series. These men are attending the conference because they will have to learn Kodály's gestural language in order to communicate with the aliens.

When the mothership lands, a bizarre acoustic and visual communication begins, associating various colored flashes with the different sounds. The technicians working on the the scientific installations begin to reproduce the five sounds using an analog synthesizer, the ARP 2500, played by the technician Jean Claude. In conversation with Bob Balaban, the actor Phil Dodds explains that this instrument "*is a little like a Moog and an electronic piano. It has a keyboard, but by pulling a lot of dials and switches, the machine can make any sound imaginable, since the sound is produced by electronic circuits, not by reeds or strings or anything like that*" [5]. Spielberg noticed Dodds' ability to use ARP. He was impressed and asked him if he wanted to be in the film, to which he agreed.

The beginning of the "learning" sequence, when communication between humans and aliens first gets started, is marked by two quick exchanges between the technicians which take place at two distinct times: the first when the first spaceships appear and the second with the mother ship:

- *Ok. Start with the tone.*
- *Tang. Go.*
- *Up a full tone.*
- *Down a major third.*
- *Now drop an octave.*
- *Cool blue. Go.*
- *Up a perfect fifth.*
- *Nothing. Nothing at all.*
- *Give me a tone.*
- *"Re" to the second. Up a full tone.*
- *"Mi" to the third. Down a major third.*
- *"Do" to the first. Drop an octave.*
- *"Do" up a perfect fifth.*
- *"Sol" to the fifth.*

The dialogue is clear and highlights the need to understand that it is really a mathematical musical language. Expressing in words the relationships between intervals is somewhat like retracing the path followed by Pythagoras to understand the mathematical nature of sounds and their numerical relationships.

The second dialogue is more precise in its use of musical terms:

- *If everything's ready here on the dark side of the moon, play the five tones.*
- *Give her six quavers, then pause.*
- *She sent us four quavers, a group of five quavers, a group of four semi-quavers …*
- *What are we saying to each other?*
- *It seems they're trying to teach us a basic tonal vocabulary.*
- *It's the first day of school, fellas. Take everything from the lady. Follow her pattern note for note.*
- *We have a translation airlock on their audio signal. We're taking over this conversation now.*

From the dialogue, it is clear that the language used is purely musical and not mathematical or physical. Moreover, in one of the scenes, the five tones have not yet been fully exposed, but the base plays D, E, C, and the spaceship responds C, G. This is a clear example of completing a phrase, and if we consider that the last sound of the five tones is G, which is the fifth of the C scale, it is not difficult to understand that the series remains incomplete on what in harmony is called the dominant and that this leaves the musical discourse open. Williams' intention is precisely to create an opening for humans and aliens to dialogue.

The mother ship begins to perform a sequence of notes that actually appears as a complete piece of music involving imagination and creative inspiration. This is *The Conversation,* written for 3 oboes, double bass, and tuba. It is a genuine dialogue between the instruments, presented in atonal form, developing the series of five notes. The oboes represent humans speaking while the tuba and double bass represent the aliens speaking. When the conversation comes to an end and the mothership opens, it returns the abductees, including Barry. One of the aliens turns towards Lacombe and the two begin to communicate using the gestural system devised by Kodály to identify the five tones. In this the first ever interplanetary greeting, sound, gesture, and meaning are woven together as Spielberg had imagined things. Roy is one of the humans who hears the "call." A small alien takes him by the hand and leads him inside the spaceship for his journey of discovery.

Close Encounters represents our human capacity to make contact with the cosmos and with other civilizations through music. Although this is a cinematographic reality, it is also true that the choice of using music to communicate is not a customary one. Spielberg could have chosen any other musical sequence, a melody, or even a piece of music, but instead he chose notes in the Pythagorean ratios of the tetraktys and non-verbal communication was achieved through chironomy, thereby linking Guido D'Arezzo with Zoltan

Kodály. In a certain sense, *Close Encounters* represents a kind of digest of the research for a Musica Mundana from Pythagoras to gravitational waves.

References

1. S. Spielberg, *Close Encounters of the Third Kind,* Film, 1977
2. I. Xenakis, *Musica Architettura* (Spirali, Milan, 1982)
3. D. Huckvale, The Occult Arts of Music (McFarland & Company, Inc, Publishing, London, 2013)
4. C. Casini, *Storia della Musica, dall'antichità classica al novecento* (Bompiani, Milan, 2012)
5. B. Balaban, Close Encounters of the Third Kind Diary (Paradise Press, Inc., 1978)

12

Conclusion

He says the Sun came out last night…he says it sang to him. [1].

Steven Spielberg's film is a concentrate of what is meant by the music of the spheres; it seems to sum up all the philosophical, anthropological, scientific, and musical inquiry pursued over the centuries, kindled by our need to understand what music is and why it is such an integral part of us. But the musicological point of view still suffers from a philosophical and psychological justification that does not lead to just one truth, but rather to a multiplicity of truths that can be proven only on small scales, but certainly cannot claim to be universal.

This justification has been constructed over the centuries from the need to answer existential questions such as "Why does man make music?", "Where does music come from?", and "What is music?" These questions have been answered in a variety of ways, from attempts by Darwin to psychological doctrines, but they all start from reasoning based on the senses and our perception; we are a long way from the method proposed by Descartes. Moreover, while certain statements may be true for Western music, the same may not apply to the music of other peoples, and so there is no reason to think that it will apply to alien peoples.

According to Newton, as reported in [2], it is only through inexperience that interpreters so frequently transform prophetic forms and expressions to justify whatever their fancy and speculations suggest to them, and according to Santillana, *"it takes a specialized mathematician to translate Archimedes"* [2]. What appears to be true is given by the mathematical basis of music, and I am not referring to the equal temperament structure, but to that Pythagorean part built on the basis of simple mathematical ratios and recognizable in any natural context. According to Xenakis, *"music is the harmony of the world, but homomorphized by the domain of current thought. This means that music has risen to the driving levels of pure mathematics, which has probed the abstract rudiments of notions and physics that can now plunge deep down into the abyss of the protean exchanges of matter"* [3].

Astrophysicists claim that alien civilizations could communicate with us using the frequency band of neutral hydrogen (1420.4 Hz) in the 21-cm spectral line. This is an easily detectable frequency, and since the universe is rich in hydrogen, it may well be the best way to establish a communication channel. However, there is also a musical method. The one shown in Spielberg's film is a good example, and it has actually already been widely used, as evidenced by the fact that sound recordings have been launched into space and music has been sent by electromagnetic waves.

The question of what music is still remains open. But I believe the answer has already been formulated: the ancient peoples on Earth believed in an initial creative sound, and the discovery of the cosmic background radiation and gravitational waves have shown that the ancients were right. Music thus constitutes both a material and an immaterial proof of the existence of our Universe and its birth. The world, our universe was born from a sound, a sound that we cannot hear with our ears, but we can hear using the powerful means provided by mathematics and physics. Since we are an integral part of the universe, we are made of the same matter that stars are made of, namely, protons, neutrons, and electrons, so music is innate to us, and not for psychological reasons, but for purely physical ones. Atomic particles vibrate and this vibration can be likened to a sound, so our whole body sings. The need to externalize singing is therefore a result of our vibratory nature. If we were made of atoms that were devoid of any motion, we would have neither ears nor musical intelligence; we would be part of an immobile universe built upon different physical laws from our own.

Radioastronomy has taught us to listen to the cosmos. We will not hear melodic music based on our own musical system, but we will hear music based on mathematical relationships. When neutral hydrogen sings for us, it uses its own recognizable "vocal spectrum," made to measure, and so does every element that makes up the universe. When we detect all the different signals, it is as if we were listening to a symphony written by the universe itself. This is the Musica Mundana that so interested Pythagoras, Plato, Aristotle, and the neo-Pythagoreans, and it is the music they tried to explain to us through reason. Aristotle claimed that we could not hear it because we had no experience of the primordial silence, but today we are indeed able to hear it, because we have been able to detect its faint breath, in the form of the fossil radiation.

While it is true that science can thus explain what music is and where it comes from, it is also true that it puts an end to the imagination and reasoning by which we have set the criteria for using Musica Istrumentalis. The creation of musical works has often been the subject of a search for perfection and a representation of the cosmos, or the primordial order that binds all things together through harmony. In any case, the appeal of a Musica Mundana has by no means been undermined because, as *Close Encounters* shows, it has become the subject of new metaphysical speculations, formulated with the means of our time. The concept of Musica Mundana was not abandoned at the end of the seventeenth century, but is still alive today, transformed in its image and perhaps even in its function, but present in the mind of man.

In *Close Encounters*, Spielberg chooses to entrust the song to Indians. He could have chosen Europeans or Americans, but instead he chooses Indians because of the difference in musical culture and philosophical doctrines. Indians are perhaps more open to feel the vibrations of the cosmos since Indian thinking has a very different history. At the same time, they are a people with an ongoing interest in their roots, cultivating the tradition of their ancestors, hence bringing us back to the ancient creation myths. While Americans seem frightened by possible alien contact, Indians remain largely untroubled, manifesting a kind of normalcy about this that would be quite foreign to Westerners. What Westerners want to know Easterners are ready to simply trust. Moreover, the Indian song shows that a different culture, lacking a tempered system, can recognize a message from "heaven" just as a technologically advanced culture can. Only *the art of forging sounds*, to paraphrase Stravinsky, is capable of communicating without using idioms.

The film thus constitutes a bond that unites NASA, Pythagoras, Kepler, Kodály, and Spielberg, a bond that is not accidental but causal, like the search for an answer to the big questions. The celestial spheres sing a harmonic music, so Kepler's intuition was based in truth, and we may therefore say that *"the Sun sings for us."*

References

1. S. Spielberg, Close Encounters of the Third Kind, 1977
2. G. Santillana, H. Dechend, *Sirio* (Adelphi, Milan, 2020)
3. I. Xenakis, *Musique. Architecture* (Casterman, Tournai, 1971)

Appendix

Soundtrack for This Book

This appendix lists all the musical passages mentioned in the book.
  Hildegard of Bingen, *Symphonia armonie celestium revelationum*
  Orlando di Lasso, *In me transierunt*
  W.A. Mozart, *String Quintet in G minor K 516*
  W.A. Mozart, *Der holle rache* from *The Magic Flute*
  G. Bargagli, *La Pellegrina,* Intermedio No.1 *Armonia delle sfere*
  A. Archilei, *Dalle più alte sfere,* in *La Pellegrina* by G. Bargagli
  L. van Beethoven, *Sonata Op.27, No.2 "Au Clair de Lune"*
  L.van Beethoven, *Symphony No.5*
  L. van Beethoven, *Cavatina* from *String Quartet No.13, Op.130*
  L. van Beethoven, *String Quartet Op.18, No.1*
  J.S. Bach, *Brandenburg Concerto No.2 in F major*
  J.S. Bach, *Prelude and Fugue No.1* from *The Well-Tempered Clavier*
  J.S. Bach, *Gavotte en rondeau* from *Violin Partita No.3 in E major*
  Chuck Berry, *Johnny B. Goode*
  Louis Armstrong, *Melancholy Blues*
  The Beatles, *Across the Universe*
  The Beatles, *Lucy in the Sky with Diamonds*
  T. Albinoni and R. Giazotto, *Adagio in G minor*
  J. Strauss, *By the Beautiful Blue Danube*
  J. Strauss, *Also Sprach Zarathustra*

© The Author(s), under exclusive license to Springer Nature Switzerland AG 2023
M. Agrò, *Music and Astronomy*, Springer Praxis Books,
https://doi.org/10.1007/978-3-031-41524-1

Appendix

C. Debussy, *Clair de lune*, from *Suite Bergamasque*
A. Dvořák, *Song to the Moon*, from the opera *Russalka*
A. Dvořák, *Symphony No.9, "From the New World"*
R. Rodgers, L. Hart, *Blue Moon*
J. Mercer, H. Mancini, *Moon River*
B. Howard, *Fly Me to the Moon*
G. Holst, *The Planets*
B. Maderna, *Serenata per un satellite*
J. Cage, *Atlas Eclipticalis*
Pink Floyd, *Moonhead*
Pink Floyd, *Astronomy Domine*
Pink Floyd, *Interstellar Overdrive*
The Byrds, *Armstrong, Aldrin and Collins*
D. Bowie, *Space Oddity*
D. Bowie, *Starman*
D. Bowie, *Life on Mars?*
D. Bowie, *Hallo Spaceboy*
G. Ligeti, *Atmosphères*
G. Ligeti, *Lux Aeterna*
G. Ligeti, *Aventures*
G. Ligeti, *Kyrie*
J. Hendrix, *Third Stone from the Sun*
J. Hendrix, *Up from the Skies*
Rolling Stones, *2000 Light Years from Home*
Dik Dik, *Help Me*
J. Meek, *I Hear a New World*
J. Meek, *Telstar*
Kraftwerk, *Spacelab*
S. Whitman, *Indian Love Call*
Tom Jones, *It's not unusual*
G. Donizetti, *Il dolce suono* from the opera *Lucia di Lammermoor*

Index

A
Adamo, A., vii, viii, 1, 55
Aetius, 13
Albinoni, T., 65
Alcuin, 36
Aldrin, B., 71
Alpher, R., 51, 53
Anaximander, 16
Anaximenes, 13, 16
Anderson, I., 75
Apollonius of Perga, 24
Archilei, A., 93, 105
Archilei, V., 93
Archytas, 15, 18–20
Aristotle, 17, 21, 25, 31, 102
Armstrong, L., 60, 105
Armstrong, N., 71
Asimov, I., 77
Augustine, Saint, 36, 37

B
Bach, J.S., 44, 45, 48, 60, 105
Bacon, F., 38, 41
Balaban, B., 94, 97
Bardi, G., 38, 93
Bargagli, G., 93, 105
Bartók, B., 45
Beatles, The, 60, 66, 74, 105
Beethoven, L. van, 44, 45, 49, 55, 60, 65, 70, 94, 105
Benedetti, G.B., 39
Benedict XVI, Pope, 55
Berry, C., 60, 105
Blüthner, J., 48
Boethius, 35–37, 54, 69
Boltzmann, L., 45
Boulez, P., 47
Bowie, D., 72, 73, 77, 80, 106
Bradbury, R., 78
Brahe, T., 27, 29, 30, 32
Brahms, J., 44
Braun, W. von, 48
Bruckner, A., 45
Bruno, G., 38, 39, 84
Burke, B., 52
Burton, T., 82
Byrds, The, 72, 106

Index

C

Caccini, G., 39, 93
Cage, J., 71, 72, 106
Cassini, G.D., 61, 62
Censorinus, 21
Charlemagne, 36
Coleman, C., 74, 77
Collins, M., 71
Combarieu, J., 43
Copernicus, N., 27, 31, 33, 38, 52
Culbertson, F., 74, 75
Curwen, J., 96
Cutispoto, G., vii, viii

D

Dahlhaus, C., 1, 43
D'Alembert, Jean Le Rond, 40
Däniken, E. von, 8
Darwin, C., 43, 101
de' Cavalieri, E., 93
de' Medici, F., 93
de Santillana, G., 10, 101
Debussy, C., 47, 70, 106
Dechend, H. von, 10
Democritus, 55
Descartes, R., 40, 41, 43, 101
Dicke, R., 51, 52
Dik Dik, vii, 74, 80, 106
Dodds, P., 97
Donizetti, G., 84, 106
Duchovny, D., 82
Dunstable, J., 36
Dvořák, A., 70, 71, 106

E

Eddington, A.S., 64
Edison, T.A., 57
Ehman, J., 60, 61
Einstein, A., 2, 41–49, 52, 64–66
Einstein, E., 46
Einstein, M. (Maja), 48

Elton, J., 77, 79
Empedocles, 13, 22
Ende, M., 4
Epicurus, 39
Euclid, 13, 30
Eudoxus of Cnidus, 24

F

Fallaci, O., 72
Fibonacci (Leonardo of Pisa), 91, 93, 97
Foster, J., 82, 83
Friedmann, A., 52

G

Gagarin, Y., 73, 74
Galilei, G., 39
Galilei, V., 37–39
Gamow, G., 51–53
Gerst, A., 76, 78
Giazotto, A., 65
Giazotto, R., 65
Giskin, A., 46
Glover, S.A., 96
Gödel, K., 45
Goebbels, P.J., 82
Gordon, R., 72
Gould, G., 65
Greene, B., 55

H

Hadfield, C., 73, 74, 77
Händel, G.F., 47
Hanslick, E., 43, 44
Haydn, J., 46, 48
Heisenberg, W.K., 45
Helmoltz, H. von, 16, 38, 43, 44
Hendrix, J., 74, 106
Heraclitus, 13, 16
Herman, R., 51
Hermann, R., 48

Herodotus, 66
Hildegard of Bingen, 36, 105
Hindemith, P., 48
Hippasus of Metapontum, 14
Hitler, A., 82
Hobsbawm, E., 57
Holst, G., 70, 74, 106
Hoyle, F., 53
Hubble, E., 52
Huygens, C., 40

Iamblichus, 13

Jeans, J., 48, 49
Joachim, E., 44
Joachim, J., 44
John the Apostle, 6
Jones, T., 82
Judica-Cordiglia, A., 79
Judica-Cordiglia, G., 79

Kandinsky, W.W., 47
Kepler, J., 27–35, 39, 52, 92, 103
Koch, C., 76
Kodály, Z., 89, 95–98, 103
Kolosimo, P. (Pier Domenico Colosimo), 8
Kraftwerk, 75, 78
Kubrick, S., 72, 81, 82, 88
Kumar, M., 49

Lasus of Hermione, 14
Leibniz, G.W., 41, 43
Lemaître, G., 5, 52
Lévi-Strauss, C., 7, 10

Ley, W., 49
Lietz, H., 48
Ligeti, G., 72, 81, 106
Lindgren, K., 74, 76
Longuet-Higgins, H.C., 45
Lu, E.T., 74, 75
Lucas, G., 87, 88
Lucretius, 39

Maderna, B., 71
Maestlin, M., 27
Malinowski, B., 9
Malvezzi, C., 93
Marenzio, L., 93
Marić, M., 45
McCartney, P., 60
Meek, J., 74, 106
Meir, J., 76
Mersenne, M., 40
Montalbetti, P., vii, viii
Montalenti, U., 71
Morgan, A., 76
Mozart, W.A., 45, 46, 48, 60, 94, 105

Newton, I., 30, 33, 41–49, 52, 101

Ocko, B., 46
Odifreddi, P., 10, 33

Parmenides, 13
Parmitano, L., 76, 79
Peebles, J., 51, 52
Penzias, A., 51–53
Peri, J., 93
Pesquet, T., 74, 76

Petersen, W., 4
Philolaus, 17, 20, 22
Pink Floyd, 71, 73, 106
Planck, M., 44, 45
Plato, 21, 22, 25, 29, 37, 44, 102
Plutarch, 16
Ptolemy, C., 2, 13, 24, 25, 31, 37
Pythagoras, 2, 13–25, 31, 37, 38, 41, 42, 54, 84, 92, 93, 97, 99, 102, 103

R

Rameau, J.-P., 43
Ramm, H., 48
Reisig, G., 48
Reiter, T., 74, 77
Robertson, E., 74, 85
Roll, P., 51
Rolling Stones, 74, 106
Rosen, N., 64

S

Sacks, O., 82
Sagan, C., 58, 60, 82, 83
Schirra, W. Jr., 75, 78
Schneider, M., 3
Schoenberg, A., 47, 96
Schubert, F., 44, 47, 48
Schumacher, E., 45
Seidel, T., 46
Sitchin, Z., 8–10
Sommerfeld, A., 44
Spencer, H., 43
Spielberg, S., 2, 59, 84, 87, 88, 90, 93–95, 97, 98, 101–103

Stafford, T.P., 75, 78
Strauss, J. Jr., 69, 81, 105
Strauss, R., 81, 82, 92
Stravinsky, I., 40, 103
Suzuki, S., 45

T

Taft, W.H., 57
Tesla, N., 83
Thales of Miletus, 15, 16
Theon of Smyrna, 13
Thorn, K., 83
Tinctoris, J., 36
Truffaut, F., 89

W

Wagner, R.W., 46
Weber, J., 65
Welles, O., 87
Wells, H.G., 87
Whitman, S., 82, 106
Wilhelm, F., 44
Wilkinson, D., 51
Willaert, A., 36
Williams, J., 84, 90–95, 98
Wilson, R., 51–53

X

Xenakis, I., 92, 102

Z

Zarlino, G., 36–39, 43
Zemeckis, R., 82